Praise for
# The Logic of Failure

"Dietrich Dörner is an internationally respected figure in the field of cognitive behavior. What is revolutionary about Mr. Dörner's analysis is his conclusion that not understanding how our actions will affect the world around us is, in its way, almost rational. Failure has its own kind of logic...Mr. Dörner's work helps us understand how good intentions come to naught and how our minds—one wants to say our natures—substitute managing the minutiae of bureaucracy and public policy for grappling with concrete human problems....Mr. Dörner refrains from drawing political conclusions, however, hewing closely to the hard facts of his scientific research."
—*The Washington Times*

"This ingenious manual will assist problem solvers in all fields."
—*Publishers Weekly*

"Dörner, a winner of Germany's highest science prize, the Leibniz Award, makes his contribution to the study of complexity by demonstrating just how difficult and problematic decision making can be. Happily, his methodology is both elegant and revealing....The implications here are substantial, for he has created a basic blueprint for testing decision-making skills and a broad model for improving them....A provocative and important road map for years of future scientific experiment and investigation."
—*Kirkus Reviews*

"Unprecedented computer-simulated research....a fascinating read!
—Stephen Covey, author of
*The Seven Habits of Highly Successful People*

"Quick, somebody give Bill Clinton...copies of *The Logic of Failure*. Hand them to CEOs of the Business Week 1000, while you are at it. Everybody knows that people in authority make dumb mistakes. Dietrich Dörner explains why they do so, drawing on psychological experiments conducted via computer simulations with role-playing volunteers. Readers will recognize themselves or colleagues in the volunteers....Dörner believes that guided role-playing can help people understand what they are doing wrong and get better at making decisions. Writes Dörner, 'Anybody who thinks play is nothing but play and dead earnest is dead earnest hasn't understood either one.'"
—*Business Week*

"Dörner, a professor at the University of Bamberg and director of the cognitive anthropology project at the Max Planck Institute in Berlin has impeccable credentials in this special branch of cognitive psychoanthropology. And he graces us with the nicest title so far. His lively treatise, accessible to the cultivated lay reader, capitalizes on real-life cases and refined ad hoc experiments...Lucid, well-balanced, wellsupported and instructive."
—*Nature*

"One of the best management titles of the year, this is a necessary addition to both psychology and management collections of all types."
—*Library Journal*

"An especially important book that deals with the nature and origins of mistakes in a way that has no precedent."
—James Reason, author of
*Human Error*

"Plenty of humor and fascinating anecdotes in a serious yet enjoyable book."
—*The New Scientist*

# The Logic of Failure

Translated by
Rita and Robert Kimber

# The Logic of Failure

Recognizing and Avoiding Error
in Complex Situations

## Dietrich Dörner

A Merloyd Lawrence Book

A Member of the Perseus Books Group
New York

*Library of Congress Cataloging-in-Publication Data*
Dörner, Dietrich.
[Logik des Misslingens. English]
The logic of failure : recognizing and avoiding error in complex situations / Dietrich
Dörner : [translated by Rita and Robert Kimber]
p.      cm.
Originally published : New York : Metropolitan Books, 1996.
"A Merloyd Lawrence book."
Includes bibliographical references and index.
ISBN 0-201-47948-6
1. Decision-making. 2. Problem Solving 3. Planning-Psychological aspects.
I. Title.
[BF448.D6713      1997]
153.4'2—dc21                                                97-20511
                                                                CIP

Previously published in the United States by Perseus Publishing
Published in the United States by Basic Books,
A Member of the Perseus Books Group

Cover design by Suzanne Heiser
Text design by Kate Nichols

Books published by Basic Books are available at special discounts for bulk purchases
in the U.S. by corporations, institutions, and other organizations. For more information, please contact the Special Markets Department at the Perseus Books Group, 11
Cambridge Center, Cambridge, MA 02142, or call (800) 255-1514 or (617) 252-5298 or
e-mail special.markets@perseusbooks.com.

DHAD      05 06   25 24 23 22 21 20 19 18 17

Find us on the World Wide Web at
http://www.basicbooks.com

# Contents

# Acknowledgments

Anyone who has ever written a book knows that one doesn't write it alone.

I thank my wife, Sigrid, for suggesting many improvements and for providing background.

I thank my father for the idea of this book and for many discussions about it.

I am grateful to Michael Koch for suggesting the title.

I thank Lydia Kacher for her vigilant and speedy work and for her patience.

I thank Kristin Härtl for the rediscovery of the Fermat formula.

I thank all my colleagues in the psychology department of the University of Bamberg for many Thursdays.

I thank Hermann Gieselbusch of Rowohlt for his patience, his understanding, and his many suggestions and for exerting pressure in the right place and at the right time.

Translating a book, assuming it is not simply a technical manual, essentially means rewriting it. The translators in this case have, I feel, succeeded admirably at that creative task, and I would like to thank Rita and Robert Kimber for their able rendering of this text. They have managed to preserve in the translation that quality which is so hard to define and grasp: the "spirit" of the original.

I would also like to express my heartfelt appreciation to my editor at Metropolitan Books, Sara Bershtel, for the skill, patience, and care with which she has seen this difficult project through to its conclusion. My thanks as well to Diana Gilooly and Roslyn Schloss for their thorough and sensitive work on this manuscript.

And finally I would like to remember Bjela, with all gratitude.

D.D.

# The Logic of Failure

# Introduction

We were in high spirits. The physicist laughed as he told his story: "Everybody had agreed on the proposed plan. The mayor had the support of both the citizens and the city council. Because the volume of traffic downtown and the resultant noise and air pollution had become intolerable, the speed limit was lowered to twenty miles per hour and concrete "speed bumps" were installed to prevent cars from exceeding it.

"But the results were hardly what the planners anticipated. The lower speeds forced cars to travel in second rather than third gear, so they were noisier and produced more exhaust. Shopping trips that used to take only twenty minutes now took thirty, so the number of cars in the downtown area at any given time increased markedly. A disaster? No— shopping downtown became so nerve-racking that fewer and fewer people went there. So the desired result was achieved after all? Not really, for even though the volume of traffic gradually went back to its original level, the noise and air pollution remained significant. To make matters worse, during the period of increased traffic, word had gotten around that once-a-week shopping expeditions to a nearby mall on the outskirts of a neighboring town were practical and saved time. More and more people started shopping that way. To the distress of the mayor, downtown businesses that had been flourishing now teetered on the verge of bankruptcy. Tax revenues sank drastically. The master plan turned out to be a major blunder, the consequences of which will burden this community for a long time to come."

The fate of this environment-conscious town demonstrates how human planning and decision-making processes can go awry if we do not pay enough attention to possible side effects and long-term repercussions, if we apply corrective measures too aggressively or too timidly, or if we ignore premises we should have considered. Effective planning and decision making were what the physicist and his economist colleague had on their minds this pleasant summer morning as the three of us walked down a hall at the University of Bamberg. The two men had come from a large, well-known industrial enterprise, and their mission was to acquaint themselves with the computer-simulated planning games my colleagues in the psychology department and I had developed and to see whether these games might be of use to them in their company's training program. Our initial conversation was a general one about the failings of human thought and action, and implicit in it, of course, was the arrogant belief that such failings are always to be found in other people—in the mayor of a small town, for example, or in the managers of a large corporation whose business policies have brought their company to the brink of bankruptcy, or in the directors of public organizations who misdirect funds. The unspoken assumption in conversations like these is always that *we* could do much better given the opportunity.

A couple of hours later the mood had deteriorated noticeably. The two visitors had, in that period, worked through a planning game. The task was to create better living conditions for the Moros, a West African tribe of seminomads who wander from one watering hole to another in the Sahel region with their herds of cattle and also raise a little millet. Things were not going particularly well for them. Infant mortality was high and life expectancy was low; their economy was decimated by repeated famines; tsetse flies ravaged their cattle, preventing any significant increase in herd size. In short, their situation was dreadful. But now there were solutions. Money was available. Measures could be taken against the tsetse fly; deeper wells could be drilled to improve irrigation and allow an expansion of pastureland; fertilizers and the planting of different strains of millet could improve crop yield; a health service could be established. There was no end to what could be done for the Moros, at least in our computer-simulated Sahel.

The economist and the physicist went to work with a will. They gathered information, studied the map of the Moro region intently, asked questions, considered possibilities, rejected one set of plans, developed new ones, and finally reached some decisions that were fed into the computer. The machine then calculated the effects those decisions would have.

Years sped by in minutes. The computer worked like a kind of time machine. After twenty (simulated) years and two (real) hours the physicist politely, but with unmistakable irritation, called the economist's attention to the simulator's report of reduced yield from the Moros' wells. "My dear colleague, it was my opinion from the outset that all this drilling of deep wells was a bad idea, and back in year 7 of the simulation I said so in no uncertain terms."

The economist responded with barely disguised annoyance. "I don't recall that at all. On the contrary, you were still suggesting the most efficient ways to drill deep wells. And, incidentally, your ideas on health care haven't turned out to be particularly brilliant either."

The reason for this clash was the truly dismal state of the Moros, whose standard of living improved at first, only to decline again quite rapidly. In the two decades of simulation, the Moro population had doubled. Thanks to an excellent health-care system, mortality—and infant mortality in particular—had dropped sharply. Initially, too, there had been a great increase in the number of cattle, thanks to successful suppression of the tsetse fly. At the same time, the drilling of numerous deep wells made available a rich supply of groundwater that allowed the Moros to enlarge their pastureland radically. Eventually, however, the pasturage was no longer able to support the large herds, and overgrazing occurred. The hungry cattle tore up the grass roots; the vegetated land area shrank. And by year 20 hardly any cattle were left, because the pastures were almost completely barren. The drilling of still more wells, helpful in the short term, had exhausted the remaining groundwater supply all the more rapidly. The Moros were now in a hopeless situation that could only be alleviated by a massive infusion of outside aid.

How could this have happened? Our two academically trained game players were not, of course, specialists in aid to developing countries. On the other hand, they considered themselves quite capable of dealing with the given problems, and they certainly had the best intentions. Nevertheless, they made terrible decisions. They drilled wells without considering that groundwater is a resource that cannot easily be replaced. They set up an effective health-care system, reducing infant mortality and increasing lifespan, but did not institute birth-control measures. In short, they solved some immediate problems but did not think about the new problems that solving the old ones would create. They now had to feed a significantly larger population with significantly reduced resources. Everything was much more complicated than before. If no outside help had been available, the result would have been a massive famine.

It is important to note that the Moro planning game does not involve any tricks. No particular technical expertise is required to play it. Everything that happens is really quite obvious. If you drill wells, you will deplete groundwater. And if the water is not replaced (and how can groundwater be replaced in any significant amount on the southern border of the Sahara?), it will be gone. That fact is easy enough to see—in hindsight. Failures in the Moro planning game evoke dismay in the observer precisely because the cause-and-effect relationships are so simple.

No one is distressed by failing to see very subtle points that require specialized knowledge. We are distressed, however, if we overlook the obvious. And that was the case here.

The outcome of the Moro planning game illustrates the difficulties even intelligent people have in dealing with complex systems. The economist and the physicist were by no means worse planners than other people. Their actions were no different from those of "experts" in real situations.

Consider an actual disaster that occurred in southern Africa, the unintended result of a project to combat hunger in parts of the Okavango delta.[1] Scientists had outlined a simple plan: the tsetse fly would be repressed and the herds of wild animals in the region would then be replaced by beef cattle. At first everything went well. Soon, however, hundreds of additional cattle herders moved into the area. Overgrazing and drought quickly turned the originally habitable land into a desert.

Like the Moros, we face an array of closely—though often subtly—linked problems. The modern world is made up of innumerable interrelated subsystems, and we need to think in terms of these interrelations. In the past, such considerations were less important. What did the growth of Los Angeles matter to a Sacramento Valley farmer a hundred years ago? Nothing. Today, however, aqueducts running the length of the state make northern and southern Californians bitter competitors for water. Of what concern to us were religious differences in Islam forty years ago? Apparently none. The global interrelations of today make such dissension important everywhere.

It appears that, very early on, human beings developed a tendency to deal with problems on an ad hoc basis. The task at hand was to gather firewood, to drive a herd of horses into a canyon, or to build a trap for a mammoth. All these were problems of the moment and usually had no significance beyond themselves. The amount of firewood the members of a Stone Age tribe needed was no more a threat to the forest than their hunting activity was a threat to wildlife populations. Although certain animal species seem to have been overhunted and eradicated in prehistoric times, on the whole our prehistoric ancestors did not have to think beyond the situation itself. The need to see a problem embedded in the context of other problems rarely arose. For us, however, this is the rule, not the exception. Do our habits of thought measure up to the demands of thinking in systems? What errors are we prone to when we have to take side effects and long-term repercussions into account?

These questions are especially pertinent when we address such problems as environmental degradation, nuclear-weapons buildup, terrorism, and overpopulation. Like the attempt to help the Moros, efforts to deal with these dangers have created new problems or exacerbated old ones. This seeming failure of our capacity to think has prompted sweeping criticism of the human intellect, if not for the existence of our problems, then at least for our inability to solve them. The theories advanced are grandiose and run the gamut from genetic to evolutionary to culturally determined.

Some analysts complain that all our difficulties stem from the fact that we have been turned loose in the industrial age equipped with the brain of prehistoric times.[2] They see our tendency to think in simple chains of cause and effect as genetically preprogrammed and locate our inability to solve our problems in this genetic programming. Others note the conditions that evolution has placed on the development of the human cognitive apparatus.[3] The claim is that we have a strong tendency to visualize when we form hypotheses about the world and events that take place within it and that our minds therefore have great difficulty grasping problems that cannot be visualized. Still others have located the source of the trouble in male domination of society. They distinguish between "serial" male thinking and "parallel" female thinking and identify the latter as more appropriate for dealing with complex problems. Indeed, the entire tradition of Western "analytical" thinking is often blamed for all our woes.

Many popular authors have gone beyond complaint and offered sweeping cures. Some are based on mysterious new regimens of thinking and learning. Several years ago, for example, a best-selling book elucidated a method that would teach us how to think in two weeks. Another book promises to teach us "new thinking" but maintains a strict silence on what this so-called new thinking really is.[4] Many individuals and institutions publicize the benefits of courses in "creative thinking," brainstorming, synectics, the 3-W method, the Q5P method, and so on. Companies recommend (and sell) "superlearning." We can even, we are told, vastly increase our cognitive powers by learning in our sleep.

Other cures rely on facile theories about the human brain: that we use only 10 percent of our intellectual potential and we must tap into the other 90 percent; that the brain can be mapped into red, green, and white sectors and we must make greater use of the green parts of our brains than we have in the past; or that the right and the left halves of the brain have different functions and we must rely more on the right half.

What should we make of all this? The probability that there is a secret mental trick that at one stroke will enable the human mind to solve complex problems better is practically zero. It is equally unlikely that our brains have some great cache of unused potential. If such things existed, we would be using them. Nowhere in nature does a creature run around on three legs and drag along a fourth, perfectly functional but unused leg. Our brains function the way they function and not otherwise. We must make the best of that; there is no magic wand or hidden treasure that will instantly make us deep and powerful thinkers.

Real improvement can be achieved, however, if we understand the demands that problem solving places on us and the errors that we are prone to make when we attempt to meet them. Our brains are not fundamentally flawed; we have simply developed bad habits. When we fail to solve a problem, we fail because we tend to make a small mistake here, a small mistake there, and these mistakes add up. Here we have forgotten to make our goal specific enough. There we have overgeneralized. Here we have planned too elaborately, there too sketchily.

The subject of this book is the nature of our thinking when we deal with complex problems. I describe the kinds of mistakes human beings make, the blind alleys they follow down and the detours they take in attempting to cope with such problems. But I am not concerned with thinking alone, for thinking is always rooted in the total process of psychic activity. There is no thinking without emotion. We get angry, for example, when we can't solve a problem, and our anger influences our thinking. Thought is embedded in a context of feeling and affect; thought influences, and is in turn influenced by, that context.

Thought is also always rooted in values and motivations. We ordinarily think not for the sake of thinking but to achieve certain goals based on our system of values. Here possibilities for confusion arise: the conflict between treasured values and measures that are regarded as necessary can produce some curious contortions of thought—"Bombs for Peace!" The original value is twisted into its opposite. Motivations provide equally ambiguous guidelines. There are those who would say that what counts are the intentions behind our thinking, that thought plays only a serving role, helping us achieve our goals but failing to go to the root of the evils in our world. In our political environment, it would seem, we are surrounded on all sides with good intentions. But the nurturing of good intentions is an utterly undemanding mental exercise, while drafting plans to realize those worthy goals is another matter. Moreover, it is far from clear whether "good intentions plus stupidity" or "evil intentions plus intelligence" have wrought more harm in the world. People with good intentions usually have few qualms about pursuing their goals. As a result, incompetence that would otherwise have remained harmless often becomes dangerous, especially as incompetent people with good intentions rarely suffer the qualms of conscience that sometimes inhibit the doings of competent people with bad intentions. The conviction that our intentions are unquestionably good may sanctify the most questionable means.

Good intentions pursued in the name of goodness, then, are no guarantee. Our physicist and our economist were eager to construct a happy future for the Moros. The result? They set goals, they acted on them, and they failed. Why? Surely neither was responsible; nor did either fail out of shortsightedness or incomplete understanding of the situation. How could he have? After all, he had the best of intentions. It was the other guy's fault. *He* fouled up the works. That stupid idea of drilling deep wells was *his!* In the laboratory, we can undo the messes we make. In the real world, that's not so easy.

Because our thinking, with its subtle interplay of emotion and calculation, conscience and ambition, reflects the richness of the world around us, experiments to determine the characteristics of human planning and decision making in complex situations should, ideally, draw on reality. We should study a large number of actual cases—the planning and actions of real politicians, organizational directors, and corporate officers, for example. But such a project runs into difficulties. Only isolated cases are available for study, and we cannot generalize from so few examples. Furthermore, real-world decision-making processes are rarely well documented, and it is hard, if not impossible, to reconstruct them. Reports on real processes of this kind are often unintentionally distorted or even intentionally falsified.

Fortunately, computer technology allows us to simulate almost any complex situation we might wish to study, from the flora and fauna of a garden pond to the social interactions in a small city. The flexibility of computer scenarios allows psychologists and other social scientists to examine experimentally processes that were previously observable only in isolated cases. Of course, scenario situations always have the quality of a game. The situations in a computer are not real—bad administrators do not starve whole countries, and incompetent mayors are not run out of town. The fact is, however, that our participants usually took our "games" very seriously. In any case, this book provides many opportunities to reflect on what in our results should be taken in earnest and what not. Grim parallels to actual events raise the question, for example, of whether we should dismiss as a macabre joke one participant's proposal to shoot any worker whose machine turned out faulty products.

Computer simulations also enable us to observe and record the background of planning, decision-making, and evaluation processes that are usually hidden. It is easier to isolate the psychological determinants of such processes this way than it is to investigate them retrospectively in the real world. In recent years, my colleagues and I have used these computer games extensively to study problem solving by individuals and groups. In this book, I present and interpret some of our findings in order to illuminate the psychological factors bearing on human planning and decision making.

Failure does not strike like a bolt from the blue; it develops gradually according to its own logic. As we watch individuals attempt to solve problems, we will see that complicated situations seem to elicit habits of thought that set failure in motion from the beginning. From that point, the continuing complexity of the task and the growing apprehension of failure encourage methods of decision making that make failure even more likely and then inevitable.

We can learn, however. People court failure in predictable ways. Readers of this book will find many examples of confusion, misperception, shortsightedness, and the like; they will also find that the sources of these failings are often quite simple and can be eliminated without adopting a revolutionary new mode of thought. Having identified and understood these tendencies in ourselves, we will be much better problem solvers. We will be more able to start wisely, to make corrections in midcourse, and, most important, to learn from failures we did not avert. We need only apply the ample power of our minds to understanding and then breaking the logic of failure.

# One

# Some Examples

### The Lamentable Fate
### of Tanaland

Tanaland is a region in West Africa (see fig. 1). Through the middle of Tanaland flows the Owanga River, which widens out into Lake Mukwa. On the shores of Lake Mukwa is the town of Lamu, surrounded by orchards, gardens, and forests. In and around Lamu live the Tupis, an agrarian tribe. The northern and southern parts of the region are steppes. The Moros, nomadic herders who subsist on hunting and on the sheep and cattle they raise, live in the north in the area around the small town of Kiwa.

Tanaland is not a real place. It exists only in the computer, which simulates its natural features, its populations of humans and animals, and their interdependence.

We gave twelve participants in this experiment the task of promoting the well-being of Tanaland's inhabitants and of the entire region. The participants had dictatorial powers. They could carry out any measures

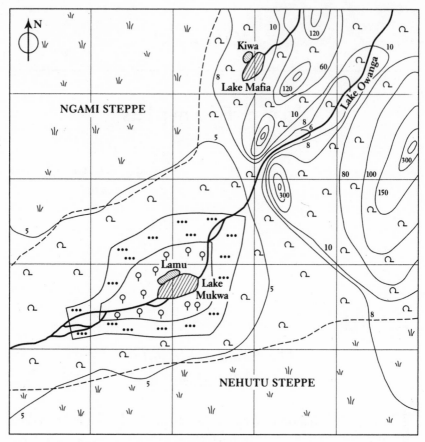

Fig. 1. Tanaland

they liked without opposition. They could impose hunting regulations, improve the fertilization of the fields and orchards, install irrigation systems, and build dams. They could electrify the entire region and mechanize it with the purchase of tractors. They could introduce birth-control measures and improve medical care. They had a total of six opportunities, scheduled at intervals of their own choosing, to gather information, plan measures, and reach decisions. With these six "planning sessions," they would determine the fate of Tanaland over a period of ten years. At each of the planning sessions, the participants could implement as many measures as they cared to. And at each new planning session, they could take into account the successes and failures of previous phases and cancel or modify earlier decisions.

Figure 2 shows the results of an average participant's governance over the ten years (or 120 months). We see that the population of the Tupis (the agrarian people) increased at first. An improved food supply and good medical care account for that. The number of children grew; the number of deaths declined. Life expectancy was higher. After the first three sessions, most participants thought they had solved Tanaland's problems. It did not occur to them that their measures had in fact set a time bomb ticking, and they were taken by complete surprise when the almost inevitable famines broke out in later years.

For our average participant in figure 2, a catastrophic famine that could not be reversed occurred in about the eighty-eighth month. It did not affect the Moro herdsmen, who had remained at a lower level of development, nearly as drastically as it did the Tupis, upon whom the blessings of artificial fertilizers and of medical care had been visited in full force. The old pattern had repeated itself: existing problems (in this case, inadequate food supply and medical care) had been solved with-

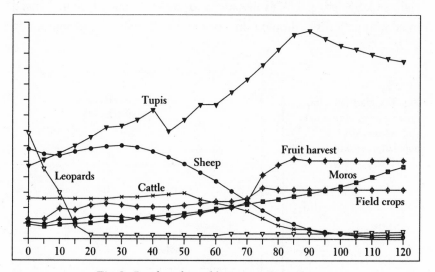

**Fig. 2.** Results achieved by an average participant
in the Tanaland experiment

out thought for the repercussions and the new problems the solutions would create.

Catastrophe was inevitable because a linear increase in the food supply was accompanied by an exponential increase in the population. Figure 3 shows the parallel development of these two factors. At first, thanks to artificial fertilizers and the deeper plowing made possible by motorization, the food supply clearly exceeded the demand. The increase in population was slower in starting, but then it quickly outran the food supply. Catastrophe was the inevitable result.[1]

Things could have gone differently. Figure 4 shows the results obtained by a different participant and suggests that a stabilization of conditions in Tanaland was possible. This participant achieved (with no little difficulty) a stable population and an overall improvement in the standard of living, results that differ dramatically from those of our average participant, whose initially very positive impact on Tanaland was followed by a disastrously negative one.

What were the reasons for success and failure? The "good" participant did not possess any expertise that the others lacked. Tanaland did not pose any problems that could be solved only with the help of specialized knowledge. The explanations for success and failure lie instead in certain patterns of thought. In a system like Tanaland's, we cannot do

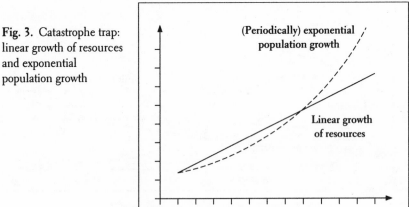

**Fig. 3.** Catastrophe trap: linear growth of resources and exponential population growth

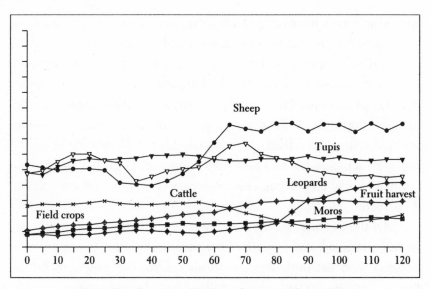

**Fig. 4.** Results achieved by the only successful participant
in the Tanaland experiment

just one thing. Whether we like it or not, whatever we do has multiple effects.

For example, one reason that Tanaland's fields and gardens are so unproductive is that mice, rats, and monkeys eat much of the crops. The obvious solution is to reduce the numbers of these pests drastically by hunting, trapping, and poisoning. Initially, extermination of the rodents and monkeys improves the yields from farms and orchards. But at the same time the decrease in small mammals brings about an increase in the insects the small mammals also feed on. And then there are the region's large predatory cats; deprived of the small mammals that are their prey, they take to feeding on cattle instead. Thus, it is possible that attempting to eliminate the rodents and monkeys will be not merely useless but actually harmful. Failure to anticipate side effects and long-term repercussions of this kind was one reason for the failures that most of our participants produced in Tanaland.

There are other reasons as well. Figure 5 compares the frequency

with which participants engaged in three types of activity: making decisions, reflecting on the overall situation and on possible courses of action, and asking questions. During the six sessions of the experiment, we
used these categories to track the thinking out loud that our participants
did; the chart shows that the relative frequency of the three activities
changed over time. At the first session, orientation activities—questioning and reflection—clearly predominate. About 56 percent of all the
recorded activities fall into these two categories. Decisions reached account for about 30 percent. Other categories account for the remaining
14 percent.

After the first session, however, the picture changes dramatically. Activities associated with analysis of the situation become fewer; those associated with decisions increase steadily. Over the course of the six
sessions, our participants clearly evolved from hesitant philosophers to
men and women of action. The participants apparently felt that their
initial questioning and reflection gave them a sufficiently accurate picture of the situation, one requiring no further correctives, whether by
gathering additional information or by reflecting analytically on results

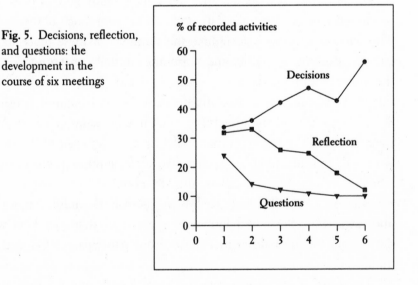

**Fig. 5.** Decisions, reflection,
and questions: the
development in the
course of six meetings

achieved. They thought, mistakenly, that they already had the knowledge they needed to cope with Tanaland's problems.

The sessions became progressively shorter, too, as measured by the average number of recorded activities for our twelve participants in each of the six sessions. At the first session, nearly fifty notations were needed to characterize the participants' behavior. In the third through fifth sessions, an average of only about thirty were necessary. The participants established their modus operandi for Tanaland in the first few sessions and did not alter it much later. Their ultimate failure shows clearly, however, that more thinking and less action would have been the better choice.

Still other factors are worth noting if we want to determine the reasons for success and failure. For some participants, the problems of the situation underwent redefinition. These participants did not consciously set out to redefine problems; the redefinitions crept up on them. One participant had decided to irrigate a large section of the Nehutu Steppe. A canal from the Owanga River would carry water to the south, where an extensive system of smaller irrigation canals would distribute the water throughout the Nehutu Steppe itself. The project required an enormous investment of money, material, and labor; and, of course, it was bedeviled by difficulties. Sometimes materials were not available on schedule; sometimes planned work quotas could not be met. The result was that the participant whose pet project this was became totally absorbed in it. When the experiment director reported that a great famine had broken out and that many inhabitants of Lamu were suffering from malnutrition or had already died from it, this participant's response was, "Yes, yes, but how is the construction of the major side canal in the Nehutu Steppe coming along?" And he never gave another thought to the famine.

An isolated case, you say. But is it in fact an isolated case? The parallels to events in the real world were so vivid here that it seemed crucial to us to study the conditions under which such things happen.

Some participants reacted cynically to repeated reports of famine. At first, the reports evoked concern, but after participants had made vain attempts to solve the problem, we began to hear remarks like "They'll just have to pull in their belts and make sacrifices for the sake of their grandchildren," "Everybody has to die sometime," "It's mostly the old and weak who are dying, and that's good for the population structure."

In the context of a game (and an electronically simulated population), we can, of course, regard these reactions as wisecracks that were not intended to be taken seriously. But here, too, the parallels with reality seem important to us: helplessness generates cynicism.

While some participants reacted with a sense of helplessness and a desire to be quit of the whole messy situation, others clearly enjoyed the "power" they had over Tanaland and assumed the role of quasi dictator with real gusto. Like a field marshal gazing off into the distance, one participant ordered fifty tractors to clear the southern forests. In his imagination, he could see the clouds of dust the fleet of tractors raised as it headed south.

Because only twelve players took part in this Tanaland game, we cannot base any generalizations on it or claim it as a genuine experiment. It did suggest to us, however, the ways that thinking, value systems, and emotions interact in decision-making processes. And that made clear to us in turn that we had to study these interlocking factors together as a whole. The parallels to real situations were obvious. Here, as in the real world, we found that our decision makers

- acted without prior analysis of the situation,
- failed to anticipate side effects and long-term repercussions,
- assumed that the absence of immediately obvious negative effects meant that correct measures had been taken,
- let overinvolvement in "projects" blind them to emerging needs and changes in the situation
- were prone to cynical reactions.

Because our game had shown such a close resemblance to reality, we were interested in examining the underlying causes of these behaviors.

## The Not-Quite-So-Lamentable
## Fate of Greenvale

Greenvale is a small town of about 3,700 inhabitants located in a
hilly region of northwest England. A municipally owned watch factory
is Greenvale's major employer, but there are, of course, other enterprises in town as well—retail stores, a bank, medical practices,
restaurants.

Greenvale is no more real than Tanaland. We simulated the main
features of this imaginary small town with a computer, thereby developing a model that would help us study thought and planning processes
in different participants. This time, however, we worked on a larger
scale and, in the course of time, made forty-eight different people mayor
of Greenvale.

Once again—unrealistically—the participants could exercise almost
dictatorial powers for ten years, during which time the citizens of
Greenvale, no matter how dissatisfied, would have no opportunity to
vote the mayor out of office. And because the region's major employer,
the watch factory, was municipally owned, the mayor could use these
powers to exert enormous influence on the economic fate of the town.
The mayor was also permitted to shape other aspects of the town's life,
such as the tax structure, to a far greater extent than any real mayor
could. In short, as in Tanaland, our participants had much more free-
dom and vastly greater possibilities to control and influence events than
anyone in the real world would ever have. We might think that this
would create the ideal basis for success, but that is not why we gave our
participants such extensive powers. Our purpose was instead to elicit
from the participants as many modes of behavior as possible. By removing the constraints of the real world, we hoped to see how people think
and act when they are entirely free to do as they wish.

So what happened in Greenvale? Some of our participants did very
well, others less so. Figure 6 shows the results that two participants we
will call Charles and Mark achieved. The graph charts the changes that

five important variables underwent over the ten years (or 120 months) of the experiment. Because the specific data are not of concern to us here, we have omitted numerical values on the vertical axis, but they are, of course, the same for both participants.

The citizens of Greenvale had every reason to be content with the administration of Mayor Charles. The town's revenue grew steadily right through to the end of the experiment, as did the production of the municipal watch factory. Unemployment never rose much above zero. The number of people looking for housing increased slightly at first but then dropped off again, so there was no crisis here. (Housing construction and rental were controlled entirely by the municipality and were therefore also the mayor's responsibility.) Given these results, the citizens' satisfaction with Charles's administration increased steadily.[2]

Mayor Mark gave Greenvale greater cause for dissatisfaction. Under his administration, the town's revenue declined steadily. After nine years, unemployment climbed off the chart. The number of people

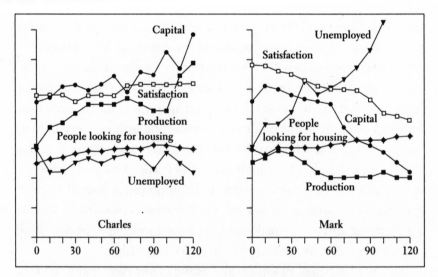

Fig. 6. Development of five critical variables
for a good (Charles) and a bad (Mark) participant in the
Greenvale experiment

without housing rose drastically. The production of the watch factory declined just as drastically. And naturally the citizens of Greenvale were dissatisfied with Mayor Mark's administration.

Charles and Mark are examples of "good" and "bad" participants. We found about equal numbers of each type among our participants and, of course, others who fell in between. What is of greatest interest to us here, however, is not success or failure but the psychology behind the results, the characteristics of thought, decision making, planning, and hypothesizing—in short, the outline of the cognitive process in our participants. If we compare the thought patterns of the people who did well in the Greenvale experiment with the patterns of those who did badly, we find very clear differences between the Charleses and the Marks.

The first obvious difference is that the good participants made more decisions than the bad ones. All participants made their decisions over the course of eight sessions. Figure 7 shows that the number of decisions increased for all participants over the first four sessions, but significantly more for the good ones. In the remaining four sessions, the difference between the two groups became considerable, as the good participants continued making more and more decisions while the bad participants started making fewer. Somehow the good participants found more possibilities for influencing Greenvale's fate.

But it was not only in the number of decisions that the participants differed. A small town like Greenvale is a complex system of interlocking economic, ecological, and political components, and—as in Tanaland—it is impossible to do just one thing alone. Any action in one area affects others. Raising taxes for a certain segment of the population, for example, may not have the single desired result of raising tax revenues. It may so alienate the affected group that they move somewhere where the tax burden is lighter. The effect of such a tax hike could well be that revenues decline rather than increase. It makes sense, then, to keep this aspect of complex systems in mind and to consider not just the primary goal of any given measure but also its potential effects on other sectors of the system.

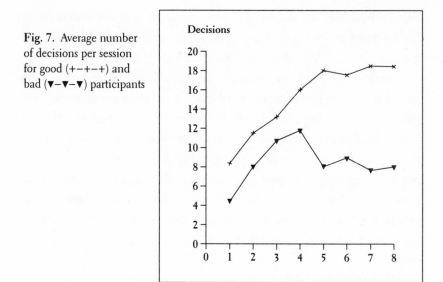

**Fig. 7.** Average number of decisions per session for good (+−+−+) and bad (▼−▼−▼) participants

As figure 8 shows, the good participants understood this better than the bad ones. We kept track of the ratio of intentions, purposes, and goals to individual decisions, and we found that the good participants reached significantly more decisions per goal. (For example, a participant could attempt to realize the goal of increased tax revenues by deciding to increase the number of jobs available in Greenvale, or she could pursue the same goal with multiple measures—increase the number of jobs, invest in product development, advertise. In the first case, we have one decision per goal; in the second, three decisions.) The good participants acted "more complexly." Their decisions took different aspects of the entire system into account, not just one aspect. This is clearly the more appropriate behavior in dealing with complicated systems.

The good and bad participants differed in the focus of their decisions, too. In figure 9, we see that the bad participants initially devoted many decisions to Greenvale's recreational offerings. Only gradually did they come around to focusing on truly important issues, such as the productivity of the watch factory, its sales, and municipal finances. The

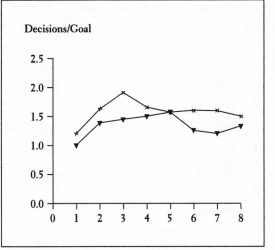

Fig. 8. The average number of decisions per goal for the good (+–+–+) and bad (▼–▼–▼) participants

good participants, by contrast, recognized early where Greenvale's real problems lay and attacked them first.

If we take a close look at thought processes—and the minutes we kept as our participants thought out loud allowed us to do just that—we find still other differences between the participants who dealt successfully with Greenvale and those who did less well. The good and bad par-

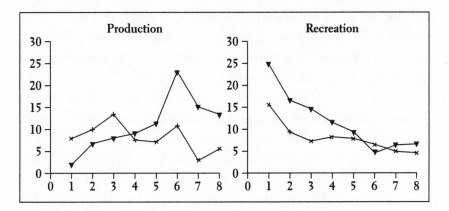

Fig. 9. Effort devoted to "increasing production" and to "recreation"
by good (+–+–+) and bad (▼–▼–▼) participants

ticipants did not differ (at least not to any statistically significant degree) in the frequency with which they developed hypotheses about the interrelation of variables in Greenvale. Both the good and the bad participants proposed with the same frequency hypotheses on what effects higher taxes, say, or an advertising campaign to promote tourism in Greenvale would have. The good participants differed from the bad ones, however, in how often they tested their hypotheses. The bad participants failed to do this. For them, to propose a hypothesis was to understand reality; testing that hypothesis was unnecessary. Instead of generating hypotheses, they generated "truths."

Then, too, the good participants asked more *why* questions (as opposed to *what* questions). They were more interested in the causal links behind events, in the causal network that made up Greenvale. The bad participants, by contrast, tended to take events at face value and to regard them as unconnected. A related finding was that good participants dug deeper in their analyses than bad ones did. A report that many young people in Greenvale were unemployed might evoke from a bad participant the reaction "That's terrible! It will certainly have a negative impact on the self-esteem of our young people. Something must be done! The director of the Youth Services Department should issue a report." A good participant's response would more likely have been "Is that so? How many are out of work? Why haven't they moved to other communities for training? How many entry-level positions do we have in our various enterprises? What career goals and interests do our young people have? Are they different for men and women?"

Given these findings, it comes as no surprise that participants who proved to be bad mayors changed the subject under discussion far more often during our sessions. Bad participants tended to leap from one subject to another, no doubt because they encountered so many difficulties in trying to solve a given problem that they dropped it and moved on. Such "slip-and-slide" transitions are typical of this kind of behavior.[3] For example, a participant working on the issue of unemployment among young people in Greenvale comes up with the idea of the municipal administration as a possible provider of training positions. Then he recalls hearing complaints that the town registration office is excessively slow in issuing new passports. And all of a sudden he is preoccupied with procedures for issuing passports and has completely forgotten about unemployment among young people.

Also characteristic of the behavior of bad participants is a high degree of "ad hocism." Even when they do not initiate the change of subject themselves, bad participants are all too ready to be distracted. A chance report on a shortage of gymnastics equipment in the municipal sports club puts an end to discussion of the difficult problem of how to increase the watch factory's sales and leads instead to a tiresome investigation of how many horizontal bars and sets of parallel bars are in the municipal gymnasium.

This instability reveals itself in certain measurable characteristics of behavior. Figure 10 shows the innovation and stability indices for good and bad participants. An innovation index indicates the degree to which the decisions made in one session deviate from those of the previous session. If a participant reaches decisions that differ totally from the ones he made in the previous session, his innovation index is high, even if he is focusing on the same set of topics as before. If his decisions resemble earlier ones, his innovation index is low. A participant's stability index is high if in a given session he operates consistently within the same range of topics as he has before. His stability index is low if he moves to a completely different set of topics.

Innovation and stability indices are not simply inverses of each other. Both may be high if a participant operates in a given session within the same range of topics as before but at the same time makes a large number of new decisions about other topics.

In figure 10 the innovation indices of the good participants are generally lower and the stability indices higher than those of the bad participants. This shows that the good participants focused their energies on the *right* fields of endeavor (they wouldn't have been successful otherwise) and that they *continued* to focus on those fields over time.

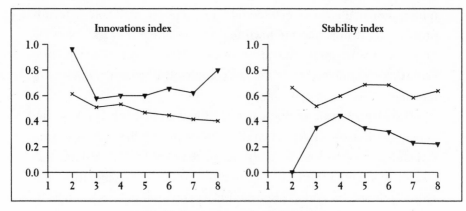

**Fig. 10.** Innovation and stability indices
of good (+–+–+) and bad (▼–▼–▼) participants

Along with aimless switching around from one field of focus to another, we also find among the bad participants just the opposite behavior—a single-minded preoccupation with one project to the exclusion of all else. One participant laboriously calculated the average distance that an average senior citizen had to walk to a telephone booth in Greenvale. The precise data he developed would provide a basis for placing new telephone booths. A project like this takes time that is then not available for other things. Of course, his interest in the social integration of the elderly is praiseworthy and testifies to the participant's social and humanitarian impulses. (Or does it?)

The good and bad participants also differed in their degree of self-organization during sessions. While the good participants often reflected on their own behavior, commented critically on it, and made efforts to modify it, the bad participants merely recapitulated their behavior. The good participants also structured their behavior to a greater degree. Their thinking out loud more frequently displayed sequences like "First I have to deal with A, then with B, but I shouldn't forget to think about C as well."

Here, too, as with the Tanaland experiment, it is worth noting that the differences between successful and unsuccessful participants were by no means limited to differences in their thought and planning processes. Confronted with stubborn problems, bad participants were frequently inclined to shift responsibility or blame onto someone else's shoulder. When they could see no other way out, their last resort was often to say: "Tom or Dick or Harry should tend to the problem." This is a normal human dodge; we see people using it every day. But it has potentially serious consequences. If, the moment something goes wrong, we no longer hold ourselves responsible but push the blame onto others, we guarantee that we remain ignorant of the real reasons for poor decisions, namely, inadequate plans and failure to anticipate the consequences.

This picture of the failings that the bad mayors of Greenvale displayed is a composite. By no means did all the participants show all these failings. Indeed, our participants—and particularly the bad ones—varied greatly in their behavior. Some were especially prone to shifting their focus from one field to another; others lacked only the ability to dig deeply enough into the problem at hand. Some participants floundered helplessly on the surface; others trapped themselves in narrow, specialized questions.

What factors shaped the behavior of our participants? The usual battery of psychological tests is useless in predicting participant behavior. We would assume that "intelligence" would determine behavior in complex situations like this, for complicated planning—formulating and carrying out of decisions—presumably places demands on what psychology has traditionally labeled "intelligence." But there is no significant correlation between scores on IQ tests and performance in the Greenvale experiment or in any other complicated problem-solving experiment.[4]

It seems likely that the capacity to tolerate uncertainty has something to do with how our participants behaved. When someone simply walks away from difficult problems or "solves" them by delegating them to others, when someone is all too ready to let new information distract him from the problem he is working on at the moment, when someone solves the problems she *can* solve rather than the ones she *ought to* solve, when someone is reluctant to reflect on his actions, it is hard not to see in such behavior a refusal to recognize one's impotence and helplessness and a tendency to seek refuge in certainty and security.

## Chernobyl in Tanaland

Up to now we have been looking at games. We have analyzed how some people found their way about in computer worlds that were in part familiar and in part alien to them. The question that arises now, of course, is what bearing the results of such games have on the real world. What does the behavior of game participants who have assumed the role of mayor of Greenvale or development dictator in Tanaland have to do with their behavior in real situations? It is highly improbable (fortunately) that our participants will ever be mayors or dictators. Are the modes of behavior our participants displayed in situations quite exotic for them reflections of *general* behavioral tendencies that we will always find if we put people into situations characterized by uncertainty, complexity, and lack of clarity? Or are these modes of behavior situation-specific, wholly a function of the exotic constellations we forced upon our participants? Here I will describe only one example. Its advantage is that it happened in the real world and not in a computer program.

On April 26, 1986, Reactor 4 of the Ukrainian atomic-energy plant in Chernobyl exploded, destroying its concrete roof (weighing thousands of tons), polluting the surrounding territory and all of Europe with radioactive particles, and intensifying the discussion of the pros and cons of atomic power, of reactor technology in the West and the East, and of whether this kind of disaster could occur again. These are certainly among the most important questions, but here I'd like to consider the immediate causes of the Chernobyl catastrophe, as psychological factors account for it entirely. It was not more sophisticated or less sophisticated technology that made the difference here but—there's no other name for it—human failure.

What happened in Chernobyl? I will not chronicle the course of the disaster in detail here but will focus instead only on key aspects that point up the significant psychological factors involved.[5]

In the Chernobyl reactor, reactivity in the core is regulated by 211 control rods. To decrease reactivity, rods are inserted; to increase reactivity, rods are withdrawn. As a safety measure, there must be at least fifteen rods inserted in the core at all times. The core of the Chernobyl reactor consists of blocks of graphite, four million pounds of them. Over sixteen hundred channels are cut through this graphite core, and through these channels run both nuclear fuel rods, which produce enormous quantities of heat, and water, which absorbs that heat, thereby cooling the core (cooling also decreases reactivity). If the heat were not absorbed, the temperature in the core would quickly rise to thousands of degrees Fahrenheit, causing an explosion or a runaway nuclear reaction called a meltdown, in which radioactive material literally melts its way through metal and concrete and into the earth. As the water absorbs heat, it boils and turns to steam. This steam, which is piped away to drive turbines and generate electrical power, is the reactor's primary product. After the steam passes through the turbines, it cools back into water and begins the cycle all over again. As long as the system is closed, radiation is contained. In addition to this primary circulating system, there is an emergency cooling system.

At the time of the accident, the reactor was due for its annual maintenance. But before undertaking the maintenance, the engineers wanted to conduct an experiment that would help them improve a safety system. The plan was to have the entire series of experiments completed before the May holidays. The engineers therefore began at 1:00 p.m. on Friday, April 25, 1986, to slow the reactor down, intending to cut it back to 25 percent of capacity. The experiments were to be conducted at this operating level. One hour later, at 2:00 p.m., the emergency cooling system was closed off from the reactor. Presumably this was done to prevent the emergency cooling system from kicking in inopportunely during the testing. Also at 2:00 p.m., however, dispatchers in Kiev requested that the reactor not be taken off the grid, because they were faced with an unexpected demand for power. Not until 11:10 p.m. that day was the reactor finally taken off the grid. The engineers began to bring it down to 25 percent of capacity so that they could run their planned tests.

But instead of dropping to the desired 25 percent of capacity, the reactor had fallen to 1 percent by 12:30 a.m. The operator had shut off the automatic controls and had tried to hit the 25 percent mark using manual controls. He did not make adequate allowance for the reactor's self-damping behavior, however, and so the reactor ultimately slowed down to 1 instead of 25 percent of capacity.

This tendency to "oversteer" is characteristic of human interaction with dynamic systems. We let ourselves be guided not by development within the system, that is, by time *differentials* between sequential stages, but by the *situation* at each stage. We regulate the *situation* and not the *process*, with the result that the inherent behavior of the system and our attempts at steering it combine to carry it beyond the desired mark. (In chapter 5, "Time Sequences," we will encounter further examples of this kind.)

Operation at low capacity is dangerous in a reactor of the Chernobyl type. In its lower ranges, it runs unevenly, as diesel engines may when idling. It operates unstably. Irregularities in nuclear fission result. Local bursts of reactivity can occur and are dangerous because they can trigger a sudden start-up of fission throughout the reactor. The operators were fully aware of these dangers and that because of them it was strictly forbidden to bring the reactor down to less than 20 percent of capacity.

The operators therefore concentrated on bringing the reactor up out of the danger zone of instability, and after half an hour they had managed to get it up to 7 percent, but only by withdrawing a large number of control rods from the core. They then decided to continue their experiment. This was probably their gravest error. After this point, the process could no longer be interrupted. The decision to continue the test program at 7 percent capacity meant that all subsequent activities would take place with the reactor in a zone of maximum instability. The operators obviously misjudged the situation. Why? Hardly because the dangers of instability had never been pointed out to them. It is more likely that the operators decided to continue the testing for several other reasons. One was probably the time pressure they were under or felt they were under. They wanted to put behind them as quickly as possible this test program that Moscow electrical engineers were demanding and that they perceived as a nuisance to begin with, and they had already been delayed almost ten hours by Kiev's request that they stay on the grid. The other reason was probably that, although the operators may well have known "theoretically" about the danger of reactor instability, they could not conceive of the danger in a concrete way. Theoretical knowledge is not the same thing as hands-on knowledge.

Another likely reason for this violation of the safety rules was that operators had frequently violated them before. But as learning theory tells us, breaking safety rules is usually reinforced, which is to say, it pays off. Its immediate consequence is only that the violator is rid of the encumbrance the rules impose and can act more freely. Safety rules are usually devised in such a way that a violator will not be instantly blown sky high, injured, or harmed in any other way but will instead find that his life is made easier. And this is precisely what leads people down the primrose path. The positive consequences of violating safety rules reinforce our tendency to violate them, so the likelihood of a disaster increases. And when one does in fact occur, the violator of safety rules may not have another chance to modify his behavior in the future.

Neglect of safety rules is by no means limited to the operators of nuclear plants like Chernobyl, Three Mile Island, and Biblis. Industrial psychologists working for chemical firms and researchers studying the causes of workplace accidents report that violation of safety rules is utterly routine. In the light of the rewards, that comes as no surprise.

But back to Chernobyl. At 1:03 a.m., shortly after the reactor had been brought to 7 percent capacity, the operators turned on the last two pumps in accordance with the experiment program. Now all eight pumps in the primary circulating system were operating, producing greater water flow and more cooling. What the operators failed to consider, however, was that this additional cooling would activate an automatic mechanism to withdraw many of the graphite rods that control the rate of nuclear fission in the reactor. Thus the system responded to excess cooling by automatically taking off its own "brakes." The operators apparently did not take note of this side effect. Completing their test program so preoccupied them that they gave it no consideration.

Another consequence of turning on all eight pumps was that the steam pressure dropped. It is obvious that if water is pumped through a heating system at an increased rate it cannot be heated as quickly as before, and a relatively lower yield of steam results. The steam yield may also be absolutely lower, and that was the case here. Because the operators needed the steam turbine for their experiment, however, they tried to offset the lowered steam production by tripling the flow of water. This in no way produced the desired result but reduced steam pressure even more. In short, it produced the opposite of the desired effect. In addition, reactivity in the core fell. The operators responded by removing even more control rods.

At 1:22 a.m. the shift foreman, satisfied with the steam level, abruptly reduced the flow of water. The water reaching the reactor core would soon be warmer; steam production would continue to increase (as would reactivity). The foreman asked for a report on the number of control rods in the reactor. There were only six to eight left, far fewer than the safety standard prescribed.

If you think that his request for a report on the number of control rods suggests an awareness of danger, you are mistaken. It was barely two minutes before the explosion, but the foreman decided to move forward with the experiment. For all intents and purposes, he was now operating a reactor without brakes.

At 1:23 a.m., the operators shut off the steam pipe that led to one of the turbines. This was a necessary step in the context of the experiment, but the consequence was that steam pressure increased further (as did reactivity). A minute later the operators attempted a kind of emergency braking. They seemed finally to have noticed something. They tried to shove the control rods back into the reactor. But that was no longer possible, because the heat that had developed in the reactor had bent the tubes into which the rods slid. At that moment, there were two explosions. The rest of the story is common knowledge.

What kind of psychology do we find here? We find a tendency, under time pressure, to apply overdoses of established measures. We find an inability to think in terms of nonlinear networks of causation rather than of chains of causation—an inability, that is, to properly assess the side effects and repercussions of one's behavior. We find an inadequate understanding of exponential development, an inability to see that a process that develops exponentially will, once it has begun, race to its conclusion with incredible speed. These are all mistakes of cognition.

The Ukrainian reactor operators were an experienced team of highly respected experts who had just won an award for keeping their reactor on the grid for long periods of uninterrupted service. The great self-confidence of this team was doubtless a contributing factor in the accident. They were no longer operating the reactor analytically but rather "intuitively." They thought they knew what they were dealing with, and they probably also thought themselves beyond the "ridiculous" safety rules devised for tyro reactor operators, not for a team of experienced professionals.

The tendency of a group of experts to reinforce one another's con-

viction that they are doing everything right, the tendency to let pressure
to conform suppress self-criticism within the group—this is what Irving
Janis identified as the great danger of "groupthink" in teams of political
decision makers such as Kennedy's advisers before the disastrous Bay of
Pigs invasion.[6]

Furthermore, the violations of the safety rules were by no means "exceptions" committed here for the first time. They had all been committed before—if not in this precise sequence—without consequences.
They had become established habits in an established routine. The operators did things this way because it was the way they had always done
them before.

We often speak of human failure in connection with the Chernobyl
disaster and with other catastrophes or close calls. There are many
meanings of the word *failure*, and surely the team of Reactor 4 at Chernobyl failed when their reactor exploded. But if we think of failure as
meaning that someone did not perform a task that he should have performed and if we look at the individual elements of behavior that ultimately produced the accident at Chernobyl, we cannot find a single
example of failure. No one who should have stayed awake fell asleep.
No one overlooked a signal that he should have seen. No one accidentally flipped a wrong switch. Everything the operators did they did consciously and apparently with the complete conviction that they were
acting properly. They did, of course, ignore the safety rules. But in doing so they did not neglect anything or do anything accidentally. Rather,
they were of the opinion that the safety rules were designed much too
narrowly for an experienced team. This conviction is not limited exclu-
sively to operators of atomic reactors. Any factory worker who cuts cor-
ners, every automobile driver who spurns a seat belt suffers from the
same flattering delusion.

In the behavior of Chernobyl's operators we find many of the traits
that characterized the participants in the Tanaland and Greenvale experiments: difficulty in managing time, difficulty in evaluating exponentially developing processes, and difficulty in assessing side effects

and long-term repercussions, that is, a tendency to think in terms of isolated cause-and-effect relationships. Against the background that the results of the Tanaland and Greenvale experiments present, the behavior of the Chernobyl operators seems quite understandable. These first-rate experts in the operation of atomic plants would have made utterly normal participants in Tanaland or Greenvale.

# Two

# The Demands

Tanaland, Greenvale, Chernobyl—when we look beyond their peculiar features, we find that they have much in common. All are
complicated systems that derive their complexity from the presence of
many interrelated variables. All are, at least in part, "intransparent": one
cannot see everything one would like to see. And all develop independent of external control, according to their own internal dynamic. In
addition, the people we saw attempt to solve problems in these systems—experiment participants or reactor operators—did not fully understand the systems; indeed, they made false assumptions about them.

The characteristics we find here—complexity, intransparence, internal dynamics, and incomplete or incorrect understanding of the system—are basic to all intricate situations in which individuals are called
upon to plan and act carefully, and they place many specific demands
on decision makers. In this chapter we will survey these demands. But
first we will take a closer look at each of these characteristics.

## Complexity

The examples we have considered so far all involve situations with a great many features. The number of cattle, the yield of the millet crop, infant mortality, the number of elderly people, the number of births . . . these elements characterize a certain state of affairs in Tanaland, and the total number of them is clearly large. The same is true of Greenvale and Chernobyl. An accurate description of these systems would only begin with the population of the town or the number of control rods in the reactor.

If we want to solve problems effectively in Tanaland, Greenvale, Chernobyl, or anywhere else, we must keep in mind not only many features but also the influences among them. *Complexity* is the label we will give to the existence of many interdependent variables in a given system. The more variables and the greater their interdependence, the greater that system's complexity. Great complexity places high demands on a planner's capacities to gather information, integrate findings, and design effective actions. The links between the variables oblige us to attend to a great many features simultaneously, and that, concomitantly, makes it impossible for us to undertake only one action in a complex system. Obviously, the supply of groundwater in Tanaland affects the millet crop. But the land area devoted to cultivating millet also influences the amount of precipitation retained in the soil. The number of training positions available in the Greenvale watch factory affects juvenile delinquency, which in turn affects the overall quality of life in Greenvale, which in turn affects population shifts into and out of the town.

A system of variables is "interrelated" if an action that affects or is meant to affect one part of the system will also always affect other parts of it. Interrelatedness guarantees that an action aimed at one variable will have side effects and long-term repercussions. A large number of variables will make it easy to overlook them.

We might think that complexity could be regarded as an objective attribute of a system. We might even think we could assign a numerical value to it, making it, for instance, the product of the number of features times the number of interrelationships. If a system had ten variables and five links between them, then its "complexity quotient," measured in this way, would be fifty. If there were no links, its complexity quotient would be zero. Such attempts to measure the complexity of a system have in fact been made.[1] But it is difficult to arrive at a satisfactory measure of complexity because the measurement should take into account not only the links themselves but also their nature. And, in any case, it is misleading (at least for our purposes here) to postulate complexity as a single concept.

Complexity is not an objective factor but a subjective one. Take, for example, the everyday activity of driving a car. For a beginner, this is a complex business. He must attend to many variables at once, and that makes driving in a busy city a hair-raising experience for him. For an experienced driver, on the other hand, this situation poses no problem at all. The main difference between these two individuals is that the experienced driver reacts to many "supersignals." For her, a traffic situation is not made up of a multitude of elements that must be interpreted individually. It is a "gestalt," just as the face of an acquaintance, instead of being a multitude of contours, surfaces, and color variations, is a "face."

Supersignals reduce complexity, collapsing a number of features into one. Consequently, complexity must be understood in terms of a specific individual and his or her supply of supersignals. We learn supersignals from experience, and our supply can differ greatly from another individual's. Therefore there can be no objective measure of complexity.

## Dynamics

Tanaland, Greenvale, and Chernobyl are all dynamic systems. They do not, like a game of chess, simply wait for the players to make moves.

They move on their own, whether the players take that movement into account or not. Reality is not passive but—to some degree—active. This fact creates time pressure. We cannot wait forever before we act, nor can we be perfectionists in our information gathering and in our planning processes. We must often make do with tentative solutions because time pressure forces us to act before we can gather complete information or outline a comprehensive plan.

The dynamics inherent in systems make it important to understand developmental tendencies. We cannot content ourselves with observing and analyzing situations at any single moment but must instead try to determine where the whole system is heading over time. For many people this proves to be an extremely difficult task.

## Intransparence

Another feature of the situations that faced our experiment participants and the Chernobyl operators is intransparence. What we really want to see may not be visible. The Chernobyl operator cannot see how many control rods are actually still in the reactor. The mayor of Greenvale cannot see the satisfaction levels of different population segments. The development director in Tanaland cannot see the current groundwater supply. Planners and decision makers may have no direct access, or indeed no access at all, to information about the situation they must address. They have to look, as it were, through frosted glass. They must make decisions affecting a system whose momentary features they can see only partially, unclearly, in blurred and shadowy outline—or possibly not at all. Intransparence thus injects another element of uncertainty into planning and decision making.

## Ignorance and Mistaken Hypotheses

If we want to operate within a complex and dynamic system, we have to know not only what its current status is but what its status will be or could be in the future, and we have to know how certain actions we take will influence the situation. For this, we need "structural knowledge," knowledge of how the variables in the system are related and how they influence one another. Ideally, this information would be available in the form of mathematical functions, though we may have to make do with a formulation as vague as the following: "If x increases, then y will decrease, and if x decreases, then y will increase." ("If unemployment increases, then the expenditures of affected households for goods that do not satisfy essential daily needs will decrease.")

The totality of such assumptions in an individual's mind—assumptions about the simple or complex links and the one-way or reciprocal influences between variables—constitute what we call that individual's "reality model." A reality model can be explicit, always available to the individual in a conscious form, or it can be implicit, with the individual himself unaware that he is operating on a certain set of assumptions and unable to articulate what those assumptions are. Implicit knowledge is quite common. We usually call it "intuition," and we say of someone who has it: "He has a feel for these things."

A good example of implicit knowledge is the kind of knowledge that enables a music lover to say, "I don't know this piece, but I know it's Mozart." She cannot say precisely what it is that identifies the piece as Mozart for her. All she can say is "It just sounds like Mozart." Another example is a doctor I once knew who was able to diagnose a certain disease with great certainty but without knowing how he did it (or rather without *being able to articulate* how he did it, for he clearly knew how). It turned out upon investigation that without being aware of it the doctor responded to the shape of the patient's lower abdomen, a certain pattern in the contraction of the musculature. Experts often display such intuition in their particular specialties.

Implicit knowledge, therefore, can be very useful. Conversely, explicit knowledge, though it can be verbalized, cannot always be made useful. For instance, a person can have theoretical knowledge and yet be unable to apply any of it in practice.

An individual's reality model can be right or wrong, complete or incomplete. As a rule it will be both incomplete and wrong, and one would do well to keep that probability in mind. But this is easier said than done. People are most inclined to insist they are right when they are wrong and when they are beset by uncertainty. (It even happens that people prefer their incorrect hypotheses to correct ones and will fight tooth and nail rather than abandon an idea that is demonstrably false.) The ability to admit ignorance or mistaken assumptions is indeed a sign of wisdom, and most individuals in the thick of complex situations are not, or not yet, wise.

People desire security. This is one of the (half) truths of psychology (for people sometimes desire insecurity too). And this desire prevents them from fully accepting the possibility that their assumptions may be wrong or incomplete. We concluded, on the basis of the Greenvale experiment, that the ability to make allowances for incomplete and incorrect information and hypotheses is an important requirement for dealing with complex situations. This ability does not appear to come naturally, however. One must therefore learn to cultivate it.

If we want to capture this chapter in a visual image, we could liken a decision maker in a complex situation to a chess player whose set has many more than the normal number of pieces, several dozen, say. Furthermore, these chessmen are all linked to each other by rubber bands, so that the player cannot move just one figure alone. Also, his men and his opponent's men can move on their own and in accordance with rules the player does not fully understand or about which he has mistaken assumptions. And, to top things off, some of his own and his opponent's men are surrounded by a fog that obscures their identity.

## Steps in Planning and Action

Having identified the general characteristics of complex situations, we can consider guidelines for coping with such situations, for we are prepared to answer two important questions: What specifically do we have to do to assess a complex problem? What demands does solving such problems place on us?

We know that we must find a way to keep track of complicated interrelations and deduce developmental tendencies, sometimes armed with only sketchy information. Beyond this, we must decide what we want to achieve and how. Later we must judge whether we have succeeded. I focus here on isolating the individual steps of such deliberations in order to study how people think and act. Figure 11 shows a possible schema for the entire problem-solving process.

Defining goals is the first step in dealing with a complex problem, for it is not immediately obvious in every situation what it is we really want to achieve. If our task is to improve the quality of life in a suburb, we should first ask what we mean by a "better quality of life." Better transportation links with downtown? Better recreational facilities? Better shopping? Improved schools? More contact between residents? The concept of a "better quality of life" could embrace all these things and many more. The goals in this situation are therefore not obvious. The only point of clarity is that the situation in this suburb is somehow felt to be inadequate. Stating a goal comparatively ("better transportation network" or "more

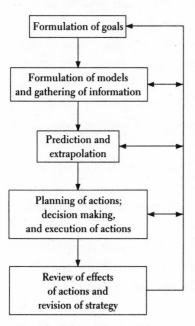

Fig. 11. Steps in the organization of complex action

user-friendly") often indicates that we don't know precisely what we want. We need to have clear goals in mind before we start forming judgments and arriving at decisions, however. Clear goals will give us guidelines and criteria for assessing the appropriateness or inappropriateness of measures we might propose.

Developing a model and gathering information follow the statement of goals. This sounds self-evident—of course we need information in order to solve a complex problem. But once again, this is easier said than done, for we must often make judgments or decisions under a deadline that allows us little time to gather all the truly necessary information.

Can we as citizens ever have a *complete* understanding of the issues on which we are asked to pass judgment on Election Day? We would have to spend all our time reading, studying, and reflecting to reach sound decisions about nuclear power, military spending, immigration, economic policy, health-care reform, and so on and so on. No one can do that. We have to work, eat, and sleep sometime, as well. And it is not just the normal citizen who lacks time to gather information. Politicians faced with the need to make a decision will rarely have time to digest even readily available information, much less to pursue new lines of inquiry.

We are constantly deciding how much information is enough. Are there any reasonable guidelines that can help us distinguish between situations in which rough-cut information will do and those in which we need greater detail? When is a general grasp of salient points adequate? When do we need to bring in a microscope?

We need, of course, to do more with information than simply gather it. We need to arrange it into an overall picture, a model of the reality we are dealing with. Formless collections of data about random aspects of a situation merely add to the situation's impenetrability and are no aid to decision making. We need a cohesive picture that lets us determine what is important and what unimportant, what belongs together and what does not—in short, that tells us what our information *means*.

This kind of "structural knowledge" will allow us to find order in apparent chaos.

Prognosis and extrapolation is the third step. Once we have acquired enough information about a situation and have formed a model that fits this information together, we should be in a position to assess not only the status quo but also developments likely to follow from the current situation. This is how things look now; what can we expect to happen next? The answer to this question is usually more important for planning future measures than is the current situation. Suppose I have a hundred dollars in my wallet now; if I had two hundred yesterday and three hundred the day before, the implications are very different than if I had fifty dollars yesterday and twenty the day before. Whether a particular economic state occurs within an upswing or a downswing is more important than the state itself. Thinking in terms of developmental trends, and understanding such trends, will enable us to prepare for future events.

Once we have a picture of the present situation and ideas about how it will change, our next step is to consider measures to achieve our goals. What should we do? Should we do anything at all? The answer is often simple-minded: we'll do what we've always done before. Acting in a ritualized way has its advantages: we don't have to start from scratch in each situation to find what the best course of action might be. This is probably why we tend to follow ritual, and it is often a reasonable choice.

On the other hand, "methodism," as Carl von Clausewitz (1780–1831) called this tendency, can impose a crippling conservatism on our activity. Many psychological experiments have demonstrated how people's range of action is limited by their tendency to act in accordance with preestablished patterns. To be successful, a planner must know when to follow established practice and when to strike out in a new direction. Recognizing the strategy appropriate to a particular situation—whether methodism or experimentation or some hybrid of the two—will help us plan more effectively.

Decisions follow planning. There are often several alternatives for action, all of which may seem good at first glance. We must decide which are in fact the best. This is rarely an easy task.

Action follows decisions. Plans must be translated into reality. This, too, is a difficult enterprise, one that calls for constant self-observation and critique. Is what I expected to happen actually happening? Were the premises for my actions correct, or do I have go back to an earlier phase of the planning process and retool? Do I have to gather information again because the information I based my actions on appears to be wrong? Do I have to develop new courses of action because the ones I have chosen are ineffective? Do I have to revise my whole model of the situation? We must be prepared to acknowledge that a solution is not working.

It is unwise, however, to abandon an established course of action too soon. Persistence often pays off, and taking the middle path between clinging stubbornly to a doomed plan and giving up a fundamentally good one at the first sign of difficulty is not easy. Finding this path, though, will give us greater chances for success.

These steps give a rough outline of the organization of complex actions. In actuality, of course, the process is not the simple progression from step to step shown in figure 11. Normally we do not just set our goals, gather information, anticipate future effects, plan measures, reach decisions, and, finally, conduct an ongoing assessment of our actions. More often we notice as we are gathering information that our goals are not formulated clearly enough to tell us precisely what kind of information we need. Or we may not notice until we have reached the planning stage that the information we had previously thought adequate for our purposes is not adequate at all. Or a well-planned measure may turn out to be completely impractical when we try to carry it out in the real world. As figure 11 indicates, however, the way back from any step to any other is generally open, and actual planning processes can involve frequent leaps back and forth between steps.

Figure 11 is a schematization, not a representation of how real people address real problems. The five steps in the diagram are a possible and, I think, helpful division of the different demands placed on people who want to act and plan effectively in complex situations. The steps contain the problems that need to be solved. In the following chapters, I describe more precisely how those problems are most effectively solved and how people in fact try to cope with them.

# *Three*

# Setting Goals

## *Requirements of Goal Setting*

Why do we formulate goals? Well, if we didn't have goals, we wouldn't do anything. No one cooks a meal, reads a book, or writes a letter without having a reason, or several reasons, for doing so. We want to accomplish some end with our activity or we want to prevent or avoid some end. We want to make something the way it "should" be or we want to prevent something that is already as it should be from changing. These desires are beacons for our actions; they tell us which way to go. When formalized into goals, they play an important role in problem solving.

We have touched on the part goal setting plays in the overall process of planning and acting. Now let's take a closer look at the pitfalls and difficulties that interfere with successful goal setting.

Goals come in many forms. I have already introduced two different kinds: positive goals and negative goals. Sometimes we act to bring about certain conditions we consider desirable, and sometimes we act to change, abolish, or avoid conditions we consider undesirable. To work toward a desirable state of affairs is a positive goal; to correct or prevent a deficient state of affairs is a negative goal.

This distinction between positive and negative goals may sound academic, but it is important. With a positive goal we want to achieve some definite condition. With a negative goal we want some condition *not* to exist. With a negative goal what it is I actually *want* is less clearly defined than with a positive goal. Negative goals (intentions to avoid something) are therefore often defined in quite vague, general terms: things have to change "somehow"; the present state of affairs, at any rate, is intolerable. Positive goals can be defined generally, too: "I need *something* to eat," for example. But it is inherent in the logic of "not" that negative goals are more likely to be vaguely defined. A "nonstove" or "nonchair" is more difficult to define than a "stove" or "chair" (though no less easy to recognize — thus a negative goal is not necessarily unclear).

"Whether things will be better if they are different I do not know, but that they will have to be different if they are to become better, that I do know," said the Enlightenment aphorist Georg Christoph Lichtenberg. In effect he was pointing up the vagueness of negative goals and at the same time warning us to be cautious in our approach to them.

We can also distinguish between general and specific goals. A general goal is one that is defined by a single criterion or by a few. A specific goal is defined by many criteria; it can be described and conceptualized very precisely. In chess, for example, a general goal is to checkmate your opponent's king. Whether a given situation on the chessboard is a checkmate or not is easy to determine, but there are a great many checkmate situations possible. Therefore, the criterion of checkmate leaves the goal of achieving checkmate only vaguely defined.

We should distinguish between a general goal and an unclear one. Unclear goals are ones that lack a criterion by which we can decide with certainty whether the goal has been achieved. "We have to make the library more user-friendly," "I want to make my room more comfortable," "We have to make the city more hospitable to pedestrians." These goals are vague and unclear. The comparatives suggest that the speaker does not know precisely what the desired state of affairs should be. All he knows is that it should be "different" from the present state.

Along with their lack of clarity—or, more precisely, within their lack of clarity—the goals quoted above reveal that they are really multiple goals. A user-friendly library is not just a library with long lending periods, or one with hours convenient for working people, or one with comfortable chairs, or one with a large selection of magazines. All these features together, however, do bring a library closer to the ideal of user-friendliness.

Pursuing multiple goals means that we have to attend to many factors and satisfy several criteria at once when we act. In addition, the interrelationship of variables in a system brings with it an interrelationship among goals. Criteria for our goals can be linked in different ways. They may be positively linked: If A is the case, then B will (usually) be the case too. A modern apartment is usually easy to heat. They may be negatively linked: If A is the case, then B will (usually) *not* be the case. A modern apartment that is easy to heat and in a good location is usually *not* cheap. And finally, they may not be linked at all; the variables bearing on them may be independent of one another.

The interdependence of variables can take different forms too. They may be directly dependent, with one variable influencing another, the other influencing the first, or both influencing each other. Or both may depend on a third variable and have no direct connection. If an apartment with modern furnishings is usually also easy to heat, this is not because modernity of furnishings has any influence on the efficacy of heating systems or vice versa. The underlying explanation for both these features is that people who can afford modern furnishings can usually also afford a well-heated, well-insulated apartment. Both variables here depend on a third.

If two goal criteria are positively linked, this simplifies things, for if we satisfy one criterion, we will necessarily satisfy the other. This is not the case when two criteria are negatively linked, for then, if we satisfy one criterion, we will necessarily fail to satisfy the other (and vice versa if the dependency is reciprocal). If we want to buy a cheap building lot, chances are that the surroundings will not be very attractive.

It is important to understand the links among goal criteria. As we have seen, in complex situations we cannot do only one thing. Similarly, we cannot pursue only one goal. If we try to, we may unintentionally create new problems. We may believe that we have been pursuing a single goal until we reach it and then realize—with amazement, annoyance, and horror—that in ridding ourselves of one plague we have created perhaps two others in different areas. There are, in other words, "implicit" goals that we may not at first take into account at all and may not even know we are pursuing. To take a simple example, if we ask someone who is healthy about her goals, she will not normally name "health" as one of them. It is, nevertheless, an implicit goal, for if we were to raise this point specifically, she would agree that maintaining her health is important. In general, however, health will become one of her explicit goals only if she falls ill.

This may sound obvious, but as we shall see, the fact that most people's actions are driven by an excessive (or exclusive) preoccupation with explicit goals accounts for a great deal of bad planning and counterproductive behavior. People concern themselves with the problems they have, not the ones they don't have (yet). Consequently, they tend to overlook the possibility that solving a problem in area A may create one in area B.

To summarize, goals may be:

- positive or negative
- general or specific
- clear or unclear
- simple or multiple
- implicit or explicit

In setting goals, we must understand these characteristics and know how to manage them. We can often replace one type of goal with another of a different type. For example, an unclear goal can sometimes be clarified into several clear goals. Or an implicit goal can be made explicit. It's also true, though, that, while some strategies of this sort have broad application, others should be applied only in particular situations.

When possible, we should try to convert negative goals into positive goals. To want to avoid something, to want to make a given situation "different"—these goals lack specificity and are inadequate as guideposts for planning and action. By virtue of its origins—a desire *not* to have something—a negative goal is often too general.

We might deduce from this that we should always try to make general goals specific. This seems reasonable. One can't play chess well if one has only the general goal of checkmating one's opponent's king. We must have specific goals as the basis for our plans and actions. It's easy to say, "Make your goals concrete," but it's not easy to do so appropriately. Take chess again: should a player, even before the opening move, set a specific goal that will guide his strategy for the whole game? "I want his king on H-1 and my queen on D-2, protected by a bishop on G-3. Furthermore. . . . Then I'll have him in check."

This goal would be very specific, but it would be foolish to adopt it in the early stages of a game, and probably no one ever would. Who knows how the game will develop? Neither player shapes the game alone; there is always an opponent. A player must be ready to seize opportunities as they emerge, and a rigid definition of final goals too early in the game can blind him to the course of developments and limit his flexibility.

So is it better not to translate general goals into specific ones? No— if particular actions are not informed by an overall conception, behavior will respond only to the demands of the moment.

One way out of this dilemma is to set intermediate goals according to the criterion of maximum "efficiency diversity."[1] A situation is characterized by high efficiency diversity if it offers many *different possibili-*

*ties* ("diversity") for actions that have a high *probability of success* ("efficiency"). In chess, examples of such situations are control over the center of the board, more men, and strategic placement of pawns. We can pursue situations of efficiency diversity even when we cannot specify our final goal. (Setting and achieving intermediate goals also has dangers, but we will examine these later.)

What should we do about unclear goals? With a general goal we always have at least one criterion for success—in chess, for example, we can verify whether we have checkmated our opponent's king. With an unclear goal, we lack even that. As we observed for "user-friendly libraries," unclear goals often contain multiple goals hidden within them. What is a "comfortable" room? Or what, for that matter, is a tax policy "favorable to labor"? What is really meant when world leaders talk about "ensuring the peace"?

"User-friendliness," "comfort," "favorability to labor," "ensuring the peace"—these are all conceptualizations, and when we have a concept, we are inclined to think that there must be something that underlies that concept, some *one* thing. But the concepts above do not denote one thing. They are complex creations that can denote a multitude of different elements and processes. A user-friendly library, we have seen, can have any of a long list of "user-friendly" characteristics. "Ensuring the peace" can mean several different things: an arms buildup, or disarmament, or a bit of a buildup at first to demonstrate one's capabilities, followed by some disarmament to demonstrate one's goodwill. Depending on the situation, any one of these options may help to ensure peace.

If we want to dispel the unclarity inherent in these complex concepts, we must "deconstruct" them. We have to take them apart and isolate what we mean *in detail* when we talk about comfort, favorability to labor, and so forth. That brings clarity. It also brings difficulties, for we will often note after we have analyzed a complex concept this way that it has no single "center" but involves many different things in different places at different times. Only after such an analysis do we see that we are not dealing with one problem at all but with a bundle of problems that may be hopelessly ensnarled, so that solving one problem may well aggravate another.

By labeling a bundle of problems with a single conceptual label, we make dealing with that problem easier—provided we're not interested in solving it. Phrases like "urgently needed measures for combating unemployment" roll easily off the tongue if we don't have to *do* anything about unemployment. A simple label can't make the complex nature of a problem go away, but it can so obscure complexity that we lose sight of it. And that, of course, we find a great relief.

By translating an unclear goal into a clear one, we often discover a multifaceted problem, one that consists of many partial problems. How do we deal with a multifaceted problem? What do we do if we are mayors of Greenvale and have to manage the town's finances (which are dismal), restructure the bureau of records (where people do nothing but fight turf battles), attract new industry, decide whether the proposed multiple-use building should be smaller so that the money saved could be spent over here where . . .

There are many ways of tackling multiple problems, but the one thing we usually cannot do is solve all the problems at once. We need to find a way to organize the list of problems. There are various possibilities:

- We can study the list to determine interdependencies. Sometimes we will find central problems that bear on a number of peripheral problems. Obtaining enough money is often a central problem, and if it is not solved, other problems cannot be solved either. Lack of money, then, will usually be an obvious central problem. Other central problems will not be so obvious. For example, that the various emotional and physical problems a young man is having can ultimately be traced to conflict in his marriage may not be obvious, particularly not to the young man himself. If we find the central problems in a situation, it is clear where we need to invest our major efforts. We can focus on solving those central problems.

- If we cannot distinguish adequately or if we cannot distinguish at all between central and peripheral problems, we can often rank problems in terms of importance and urgency. In a program to improve the recreational offerings of a community, providing a meeting place for coin collectors will be less important to the solution of the problem than will the construction of athletic facilities. The urgency of a problem is a function of the given time frame. A task that must be completed by noon today is more urgent than one that does not have to be finished until this evening. Conflicts between importance and urgency often arise. Is it better to do, by a certain time, something that is important but not urgent or something that is urgent but not so important? Becoming conscious of such conflicts is usually all we need to do to resolve them. When we are pressed for time, however, we can lose sight of the importance and urgency of the individual problems in our bundle, and then we revert to "muddling through," focusing on urgent but often unimportant problems and ignoring the truly important ones. This pattern will be familiar to just about anyone.

- Finally, we can ease the task of dealing with multiple problems by delegation. But this works only if some of the problems are independent enough of others that they can be handled in isolation over a certain period of time. We should also distinguish here between delegating problems and dumping responsibility for them onto others. The line between the two is not always easy to draw. What is the difference? Delegation means that we commission other institutions and persons to do detail work for us but that we remain conscious of the role the delegated problem has in the overall problem. We stay in touch with the delegated problem. By contrast, when we dump a problem on someone else, we instantly dismiss that problem from our mind; when it reappears in our consciousness, we respond with irritation over this new imposition on our time and attention, a sure sign that the problem will remain unsolved.

Ranking problems and delegating tasks are good strategies when we have more than one goal. Nevertheless, they do not address a basic difficulty of multiple goals. As we have observed, individual partial goals may be incompatible. In achieving one goal, we may move far from another. By solving one problem, we may make another worse.

If the solutions to two problems conflict, there are not many ways out. One is to seek a balance that yields a less-than-ideal solution to each problem. An alternative is to solve only one problem well and to forget about the other entirely.

Goals conflict with one another not by their very nature but because the variables relating to them in the system are negatively linked. Thus, a third possibility for resolving goal conflicts is to reshape the entire system in such a way that negative relationships within the system disappear. This is the strategy being used when colleges require their athletes to earn grade point averages above a certain level in order to play. These schools have taken two variables—athletic performance and academic performance—that traditionally are negatively linked and a fortiori linked them positively.

Now to implicit goals. These are dangerous because they go unnoticed at the early stages of a planning process. They emerge only after we have pursued other goals with which the implicit goals are negatively linked. For example, although many people today consider DDT an ecological plague, it was regarded as a blessing when it was first developed: at last an effective means to prevent vast insect destruction of crops, particularly in the Third World—at last an effective weapon against hunger. The problems caused by the use of DDT came to light only gradually.

By developing DDT, scientists solved one problem, but that solution caused new problems. Why did no one anticipate those new problems? The easy answer is, "Because we didn't know enough back then." But I think the lack of knowledge is secondary. More important, it seems to me, is that no one took the trouble to acquire the necessary knowledge. When we are working on a given problem, we focus on that problem alone and not on problems that don't exist yet. So the mistake is less not knowing than not wanting to know. And not wanting to know is a result not of ill will or egoism but of thinking that focuses on an immediately acute problem.

How can we avoid this pitfall? Simply by keeping in mind, whenever we undertake the solution of a problem, the features of the current situation that we want to *retain*. Simple? Apparently not.

As Brecht observed late in life, advocates of progress often have too low an opinion of what already exists. When we set out to change things, in other words, we don't pay enough attention to what we want to leave unchanged. But an analysis of what should be retained gives us our only opportunity to make implicit goals explicit and to prevent the solution of each problem from generating new problems like heads of the Hydra.

We have looked at goals and their characteristics and at the requirements for setting and managing them. Let us now look at some examples of how people in real situations deal with goals.

## General Goals and "Repair Service" Behavior

The goal we set for our Greenvale mayors was to provide for the "well-being of the citizens." This goal is of no use whatsoever as a guidepost for action. Why? Well, what does "well-being" mean? Quite a lot and quite a few different things and, therefore, without further definition, nothing at all. Well-being is the kind of unclear, complex goal I described in the previous section. It means at a minimum the securing of basic material needs: adequate food and shelter. Adequate employment opportunities are of course essential to that goal. Other components of well-being, beyond the basic ones, are medical care in case of illness, protection from crime, and a wide range of cultural offerings.

So before we start serving up well-being, we first have to identify its components as best we can. We need to break the unclear goal of well-being down into its components and take a close look both at them and at their interrelationships. And this is where many people encounter their first difficulties. They don't analyze their complex goal. They don't realize that well-being is a complex concept comprising many different components and their relationships to one another.

They accept the goal "well-being of the citizens" and go to work. Or, to put it more precisely, they start muddling through. By not breaking their complex goal down into partial goals, they almost inevitably condemn themselves to what I call "repair service" behavior. Because these mayors have no clear idea of what well-being means, they go out in search of things that are malfunctioning, and once they find them, their immediate goal becomes fixing whatever is broken.

In our Greenvale game, for instance, one mayor posted himself at the front doors of a supermarket and asked housewives what complaints they had about things in town. This sort of procedure is guaranteed to produce results. One woman complained about the dog droppings on the streets; another complained about the snail's pace at which public officials worked; still another found services for the elderly inadequate; a fourth felt that more should be done for the town library. . . .

A mayor who is guided by a randomly generated list of complaints risks giving far too much attention to relatively unimportant problems and either overlooking the truly important ones or failing to assess them properly. The upshot is repair-service behavior: the mayor solves the problems that people bring to him. This is most likely to be the case when certain ills are particularly evident. In a traffic accident, for example, the sufferings of victims with minor injuries are sometimes much more apparent than those of the severely injured because the seriously wounded are no longer screaming and therefore no longer calling attention to their plight. It may happen that those who need little help get it all and those who need it most get none. And so it often is with the partial problems of difficult and complex situations. That inadequate analysis of problem components leads to repair-service behavior is logical. How else can we deal in an unclear situation when we don't really know what it is we want?

One possible consequence of repair-service behavior is that the wrong problems are solved. Because we don't understand the connections between problems (and don't even know that we don't understand them) and have no inkling of their connection with the still ill-defined overall problem of the citizenry's well-being, we select the problems we will solve on the basis of irrelevant criteria, such as the obviousness of a problem or our competence to solve it.

One of the Greenvale mayors, for example, found himself confronted at a meeting by city administration workers demanding a pay raise. Theirs was not the most pressing problem in Greenvale, but their financial distress, combined with their close ties to the mayor as municipal employees, prompted him to drop everything else and focus all his attention on the comparatively minor problems of this group. This mayor set priorities by the criterion of obviousness.

Another mayor with some professional experience in social services found (almost with relief) that many children were having difficulties in school. She knew her way around in this field. She knew what questions to ask and what actions to take. And so she ignored other problems and devoted herself entirely to school problems, then to the problems of a certain class, and, finally, to the problems of a single individual— fourteen-year-old Peter. This mayor selected problems on the basis of her competence. She solved not the problems she needed to solve but the ones she knew how to solve.

Another possible consequence of repair-service behavior is a total disregard of failings and malfunctions that may not exist at the moment but that will emerge later. If we act on the basis of a more or less randomly generated list of complaints, we necessarily remain captives of the present moment. The implicit problems that the solution of today's problems may generate remain invisible.

Other problems that go unnoticed are those that start small, presaged only by minute signs, but develop with increasing speed. Unless we anticipate such problems, they will take us by surprise, appearing to explode out of nowhere. We should therefore take the future into account when dealing with dynamic systems. This is another principle that may sound obvious. But we will have occasion throughout this book to observe the minimal capacity and inclination people have to deal with time.

A goal that remains unclear, one that is not broken down into concrete partial goals, runs the risk of taking on a life of its own. Without concrete goals, there are no criteria that can be used to judge whether progress is in fact being made. The Tanaland dictator who became absorbed in the irrigation of the Nehutu Steppe to the exclusion of all else illustrates this point. What happened? At some point perhaps, a resident of Tanaland complained about the lack of rainfall and our participant started wondering what he could do about it. The construction of a complicated irrigation system is a challenging task fraught with difficulty, and one that can well take on its own dynamic. Human beings are so constituted that they often find challenges attractive. Casinos are evidence that people are even willing to pay for the opportunity to face challenges. The challenges must, of course, fulfill certain conditions. Success cannot be too certain or too unlikely. If success comes too easily, the game is no fun. If success is impossible, the game is too frustrating. But people find situations that offer about a fifty-fifty chance of success interesting and exciting, and they will happily focus their energies on such challenges over extended periods of time. If, in the course of constructing an irrigation system, a person experiences both successes and setbacks on the way toward completing the entire project, the task can assume its own life: the real purpose of the task—to provide improved pastureland—can disappear behind the challenge the task itself represents; indeed, it can be forgotten entirely.

Psychologist Mihaly Csikszentmihalyi coined the term "flow experience" for the fascination exerted by work that constantly poses new challenges of moderate difficulty.[2] Surgeons, chess players, mountain climbers, pilots of hang gliders, and many other individuals who engage in activities that are difficult but that also yield successes are susceptible to flow situations. A flow situation is one in which tension is built up, then released, a sequence in which the individual experiences fear of failure, triumph over obstacles, renewed fear of failure, another triumph, and so on. Inadequate concretization and elaboration of goals can leave a problem solver vulnerable to this phenomenon. An interim goal happened on by chance may seduce him into a flow situation he is helpless to escape (and perhaps may not even want to escape).

In scientific research, for which the immediate applicability of results is often not (nor should be) a criterion of success, "goal degeneration" of this kind is no minor matter. Many social scientists who have set out to write computer programs they could use to evaluate an experiment have woken up years later to find themselves computer specialists. And they will hardly have realized that they have long since lost sight of their real goal and become addicted to the fascination, challenges, and triumphs of working with a computer. An interim goal has dislodged the primary goal.

Max Horkheimer has this degenerative process in mind when he writes: "Scientists bow down to the 'concretism' of a mode of understanding that finds the works of a clock more interesting than the time the clock measures. They have all become mechanics, as it were. In their theories, they invest all their love in those things that they can deal with free of doubt. They think they can find security in things that seem absolute to them and that protect them from all contradiction. They are infatuated with neat means, methods, and techniques and pathologically underestimate, or forget, what they think themselves no longer capable of and what all of us at one time or another have hoped to achieve in the way of insight."[3]

Horkheimer has not forgotten here to offer a few hypotheses on why goals degenerate: the desire for safety and a lack of confidence in one's own capabilities. The two factors can, of course, be linked: someone completely lacking in confidence will be most intent on security.

The various phenomena that follow on inadequate definition of too general or unclear a goal develop with a peculiar logic. Inadequate analysis leads first of all to uncertainty. "For some reason" we don't know what we ought to be doing, and so we go out in search of problems. Once we find some, we must then decide which ones we will attack first. If we have no criteria based on the specification of our goals to help us set priorities, we will choose the most obvious problems—or the ones that we already know how to solve.

Not only do we then almost inevitably end up concentrating on the wrong problems but we neglect long-term considerations, especially when partial or interim goals capture our attention and displace primary goals. Realizing we are attacking the wrong problems only makes us more uncertain. How do we get out of this bind? We isolate ourselves in a task at which we feel competent, preferably one that offers both challenges and the gratification of some success.

It is essential when working with a complex, dynamic system to develop at least a provisional picture of the partial goals we want to achieve, for that will clarify what we need to do when. It is clearly not rational to pump money into improving recreational facilities if, for example, health care is underfunded. But this is precisely the kind of mistake repair-service behavior produces, as we have seen. Seizing on obvious or readily solved problems leads not to planned action but to helter-skelter responses first to one grievance, then to another.

Having roundly criticized repair-service behavior, let me conclude with a few words in its favor. The fact is that repair-service behavior is not totally irrational. It is surely preferable to correct obvious ills than to do nothing at all. The economist Charles Lindblom even recommends this kind of "muddling through" for many situations.[4] And when the philosopher Karl Popper argues for pragmatic politics driven not by lofty ideals but by the exigencies of the situation, he, too, is advocating a kind

of repair-service behavior.[5] What is important here is a proper assessment of the situation. If we can formulate goals in concrete terms, we
should; if we can't, then muddling through is better than inaction.

Ideally we should have a clear understanding of the current importance of different components in the complex we call well-being. This
knowledge will enable us not only to set ourselves the right interim goals
at the right times but also to change our priorities and move from one
goal to another. For no matter how important health care is, for example, it would be overdoing things to set up a fully equipped first-aid station in every building. Once adequate health-care facilities have been
provided, we should move on to other goals.

## Liberty, Equality, and "Voluntary Conscription"

In complex situations it is almost always essential to avoid focusing on
just one element and pursuing only one goal and instead to pursue several goals at once. In a system complicated by interrelationships, however, partial goals often stand in contradictory relation to one another.

Consider a suburban town in which the residents complain of "poor
living conditions." An analysis might show that transportation links with
the city and shopping opportunities are both inadequate. Improving
transportation into the city would be one possible goal, improving local
shopping opportunities another. But neither can be pursued in isolation
from the other. If we improve transportation, the residents of the suburb
may choose to shop in the city, where they have a much wider range of
choices than in their suburb. This will in turn lead to closings and bank-
ruptcies among local retail stores. Achieving one goal will work against
achieving the other. If transportation remains inadequate, however,
many residents of the suburb will have no choice but to shop locally.
And this will normally improve local shopping opportunities because
more retail businesses will move into the area.

In a situation like this we must weigh our options. Sometimes we will have to forget about one partial goal entirely because the other one is more important. In any case, we must consider the degrees to which we can realize each partial goal.

Contradictory goals are the rule, not the exception, in complex situations. In economic systems costs and benefits are almost always at odds. Something that doesn't cost much is rarely of much benefit, and if we want to derive great benefit we usually have to invest a lot. Because businesses want to minimize costs and maximize benefits, we have clearly found a conflict of goals. But this conflict is relatively harmless because it is one we are usually aware of. More dangerous are the situations in which the contradictory relation of partial goals is not evident. For most people who plunge naively into a planning process, the contradictory nature of better transportation links and better shopping opportunities will not be immediately obvious. Nor was it obvious in the days of the French Revolution that realizing the partial goals of liberty and equality at the same time is no easy task.

If we understand "liberty" to mean simply a condition under which there are very few constraints placed on the actions of individuals and if we understand "equality" to mean the right of equal access to the material and nonmaterial resources of a society, then "liberty" will quickly result in great inequality because those who are better equipped for certain activities (for example, those who are more intelligent) will be more successful at obtaining the resources they want, while others will be less successful. Thus, more liberty will mean greater inequality. On the other hand, the attempt to achieve a high degree of equality in a political system will produce a correspondingly high number of constraints on the individual, and that is not liberty.

It is especially difficult to reconcile conflicting goals when one of them is implicit and when we are therefore unaware of it. To return to our example of the suburban town: if the suburb offers good shopping possibilities, no one will complain about the shopping situation, because no one finds it inadequate. If, at the same time, the transportation connections to the city are poor, however, complaints will be voiced about them. If we respond to these complaints by creating better transportation links to the city, we will undercut the satisfactory shopping situation in the suburb.

Back in the late sixties, when students the world over were agitating for massive changes at their universities, the students in the psychology department of the University of Kiel, in Germany, demanded an increase in the number of slots for psychology majors. The question, raised rather diffidently by one student, as to whether there would be enough jobs for all these new graduate psychologists was greeted with loud objections. That was beside the point. The issue at hand was accommodating more students. The matter of their future prospects was deemed irrelevant and therefore not worth discussing. The point was to give everyone the opportunity to study whatever he or she chose to study.

In 1969 there was no unemployment among psychologists. At a gathering about ten years later, the entire student body demanded that the faculty ensure the creation of jobs for graduate psychologists, jobs of any kind in any place, but jobs. In the interim, unemployment among psychologists had reached relatively high levels. Now no one suffered from limits on departmental size. Now the complaint was that the large number of psychologists the universities had produced couldn't find work in their field.

Unrecognized contradictory relations between partial goals lead to actions that inevitably replace one problem with another. A vicious circle is commonly the result. By solving problem X, we create problem Y. And if the interval between the solutions is long enough that we can forget that the solution of the first problem created the second one, as in the case of unemployment among the psychologists, someone is sure to come up with an old solution for whatever the currently pressing problem is and will not realize that the old solution will create problem X again and send the circle into another cycle.

The same thing happens when current problems are so urgent that we will do anything to be rid of them. This, too, can produce a vicious circle in which we flip-flop between two problematic situations. Someone suffering from a bad headache, for example, will probably be willing to take a medication that rids him of the headache but that may, according to the label, produce a stomachache as a side effect. The headache is real; the stomachache is an abstraction, only a future possibility. To be relieved of his headache, the headache sufferer will probably be willing to take his chances. But once he has a stomachache, the situation will be reversed and he will reach for a medication that cures stomachaches, probably even if it produces headaches as a side effect.

When people recognize that they are caught in vicious circles of this kind, they find different ways to deal with them. One of these is "goal inversion." They give up one goal or even pursue the exact opposite of the original goal. The Tanaland experiment provides us with a striking example of this. After a participant had brought about marked improvement in material circumstances and medical care, famine broke out. The graph in figure 3 (p. 14) shows why: while the improvement in material circumstances followed a nearly linear progression, the population began to grow exponentially. At the same time, the starving population was hard at work on all sorts of projects. They had to build dams, dig irrigation canals, and so on. This was the occasion on which one "development director," responding to reports about the inadequate food supply, declared, "They'll just have to pull in their belts and make sacrifices for the sake of their grandchildren." The misery this participant had created he now declared a necessary transitional phase on the road to a future paradise.

Another participant went a step further in this direction. He is the one who said, "It's mostly the old and weak who are dying, and that's good for the population structure." Here the famine was not simply labeled a necessary transitional phase but was elevated to the status of a benefit. Instead of recognizing it as the catastrophe it was, this participant recast it as a desirable means of redressing the population problem.

Still another means of coping when we find ourselves pursuing or having achieved contradictory goals is "conceptual integration," or, plainly put, doublespeak. In one of our planning games, people were asked to shape a country's domestic and foreign policies. The computer simulated a background of global industrial forces, demographic structures, and climatic and ecological conditions. One participant found himself threatened on the foreign-policy front while at the same time he needed to cope with vast unemployment at home. The solution he hit on to deal with both these problems was to introduce universal military service. At the moment this idea occurred to him, however, he recalled that only a few hours earlier he had announced decisively that the government should do nothing to strengthen the military and should certainly not introduce any forced measures to that end. So our participant had created conflicting goals for himself. What did he do? He introduced *voluntary conscription*, commenting as he did so, "Everybody will surely understand the need for this."[6]

The intent behind these verbal unions of two incompatible realities seems clear to me. The perpetrator wants to keep one thing without losing the other—and *verbally* it seems to work. He may not even realize that he is trying to mix oil and water when he puts two such terms together.

It is worth noting in passing here that these verbal integrations of incompatibles can, over time, produce changes in the meanings of words. Anyone who cannot see, for example, that he ought to voluntarily accept conscription has not properly understood what *voluntary* means. "True" free will in this view can exist only if it is rooted in basic acceptance of the necessity for military service. Anyone who does not accept this necessity clearly lacks the capacity to make a voluntary decision and must therefore be forced to make that decision. In this way, terms like *voluntary* can be completely stripped of their meanings and even made to mean their exact opposites. George Orwell vividly illustrated this process with his "Newspeak" in *1984*.

Another and perhaps even more alarming form of resolving goal conflicts is by way of "conspiracy theories." We should not be held accountable for our mistakes, someone else (with evil intentions) should. The example early on of the physicist and the economist in Tanaland demonstrates this tendency. Once the catastrophe of famine was upon them, the physicist criticized the economist for drilling deep wells. He was quick to see forces at work that had undermined all his efforts to achieve positive results. And his colleague, the economist, embodied those forces. Granted, the physicist did not accuse the economist of deliberate malice, but he clearly did blame him for a lack of insight. People in other situations have come to different conclusions. In our simulations, for example, players are often inclined to blame disasters on the nature of the experiment or on the programmer. "You people set this situation up so that nobody can succeed. All you really wanted to find out was how long I'd put up with your frustrating me at every turn." This is the kind of reaction that failure often elicits from our participants.

One participant in the Greenvale experiment decided to fire the entire management staff of the watch factory and establish a workers' cooperative instead. This rather abrupt and significant change precipitated an economic disaster. Our participant could not understand why and promptly blamed it on "ill will" and "sabotage" by the workers. Everything would improve, he opined, if every worker who was caught at sabotage could be instantly shot. (In designing our game, we had not of course reckoned with anyone taking such drastic measures.) This participant later called his idea for disciplining the workforce an attempt at black humor. Maybe so, but the similarity to real-world situations is so clear that we would do well to study carefully the psychic mechanisms behind such ideas.

In my view, self-protection—the need to preserve a sense of our own competence—plays a key role here. It is difficult for us to admit to ourselves that, despite the best of intentions, we have failed. Such failures suggest that our understanding of prevailing conditions is inadequate. This inadequacy means in turn that our capability to act is limited and that we should move very cautiously. We reject that conclusion and the guilt feelings that accompany it, and so we invent conspiracy theories.

When we must deal with problems in complex systems, few things are as important as setting useful goals. If we do not formulate our goals well and understand the interactions between them, our performance will suffer. When we do not make overly general or unclear goals specific, we are likely to spend time unwisely on ineffective repair-service behavior. Ultimately, to salve our self-confidence, we may find ourselves choosing projects for their obviousness or ease rather than for their importance. If we overlook implicit contradictions among our goals, we may achieve good results initially, but in the long run we will produce many bad results. We have strategies for dealing with this phenomenon as well—goal inversion makes unintended bad results into good ones, conceptual integration erases the differences between incompatible elements, and conspiracy theories place the blame for our mistakes on others. If we can learn to recognize these tendencies in ourselves, we will be able to examine the failures in goal setting they are meant to disguise and, with work, will have a chance to improve our ability to set goals appropriately.

# *F o u r*

## Information and Models

### *Reality, Models, and*
### *Information*

The garden pool stinks. So scoop the fish out and drain the water. The bottom stinks too. So dig out the bottom and cart the stinking stuff off in a wheelbarrow. Put fresh gravel on the bottom, replant the water plants, fill the pool with water, and put the fish back in. End result: a hard day's work and two dead fish. But the pool doesn't stink anymore.

Two months later: the garden pool stinks. . . .

The sequence of actions sketched here had the goal of changing a bad situation, but the whole procedure proved futile in the end. All that work for nothing. What went wrong? The plan was thorough and the work was conscientiously performed. The mistake was made at the beginning with a badly defined goal. Changing a bad situation was equated with making the garden pool stop stinking now. This was accomplished, but the benefit was short-lived. Because of the way the goal had been defined, all effort had gone toward treating a symptom and none toward solving the underlying problem.

What are the reasons for this kind of inadequate goal definition? We can assume that an insufficient assessment of reality—a reality model that is too crude, too imprecise, or lacking altogether—is one reason. An exact and specific reality model implies knowledge about the possible interrelations within a particular system. It tells us what is important at the moment and what may be important in the future.

In the example here, the pool was deeper than it was wide and the water in it did not circulate adequately. Therefore, the lower strata of water did not receive enough oxygen, and anaerobic, foul-smelling bacteria thrived in the bottom of the pool. The proper solution would have been to install a small pump that would keep the water in motion and aerate it.

We can't focus only on what is wrong and what we want to correct. In the case of an evil-smelling garden pool we have to consider the different components that make up this particular pool and, perhaps more important still, how these components interact. We have to see the pool as a system: The water in the pool affects the fish. The waste matter of the fish affects the bottom of the pool. The condition of the pool bottom is in turn crucial for the water plants. The water plants affect the oxygen content of the water. The amount and distribution of the oxygen in the water affect the condition of both the water plants and the fish.

Perfectly clear and utterly obvious, isn't it? Unfortunately, not at all. If we look at the history of technology and industrialization or of aid to developing countries or of municipal and regional planning, we find multiple examples of the ways in which the systemic nature of the situations was ignored.

Egypt lacked sufficient electricity, so the Egyptians built the Aswan hydroelectric dam. The power from the project allowed them to supply new industries with electricity and create new jobs. But no one anticipated the side effects the dam would have. The Nile below the dam no longer carried mud with it or flooded the fields. Because this source of natural fertilization was removed, increased use of chemical fertilizers was required. That raised the costs of agricultural production and aggravated water pollution. The clear water below the dam was able to carry more silt and accelerated erosion of the river banks. Because it also carried fewer nutrients, it provided less food for the marine life in the waters off the Nile delta, with far-reaching consequences. No one thought about any of these possibilities when the dam was being planned. I could go on citing similar examples, but I won't because many have already been amply documented.[1]

What precisely is a system? A system is a network of many variables in causal relationships to one another. Within a system, a variable may even have a causal relationship to itself, as it were. The population of fish in a garden pool, for example, is partially determined by the population's own birth and death rates.

Figure 12 shows the interdependencies between individual variables in a garden pool. The arrows marked "+" indicate "positive" effects: "the more of this, the more of that" and "the less of this, the less of that." The

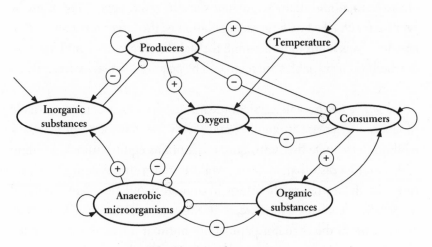

Fig. 12. A pond as a system

arrows marked "−" mean "the more of this, the less of that" and "the less of this, the more of that." The lines with circles on their ends indicate limitations; that is, the variable in question can increase only to the extent that the variables influencing it are present. The arrows on the peripheries indicate influences from outside the system or within an individual variable. We will not examine this particular system in detail, but figure 12 makes it clear that there is nothing here we could change without influencing everything else.

Thus, it is usually wise when correcting a deficiency to consider it within the context of its system. If we don't, we may treat only the symptoms and not the source of the trouble. We may also overlook the unpleasant side effects of our actions and do more harm than good in the long run. Considering the system means more than simply acknowledging the existence of many variables, however. It means recognizing the different ways the variables can affect one another and themselves. These interrelationships can be grouped into the categories of positive feedback, negative feedback, buffering, critical variables, and indicator variables.

*Positive feedback* in a system means that an increase in a given variable produces a further increase in that variable; a decline, a further decline. Animal and plant populations are to some degree regulated by positive feedback. The larger a population is, the larger it becomes. Positive feedback tends to undermine the stability of a system, and a system in which many variables are regulated by positive feedback can easily go haywire.

*Negative feedback* in a system means that an increase in one variable produces a decrease in another and vice versa. This kind of feedback tends to perpetuate the status quo. It maintains equilibrium in a system and, should a disturbance occur, works to return the system to equilibrium. Predator-prey relationships in animal populations are regulated by negative feedback. An increase in the prey population will produce an increase in the predator population, but that in turn produces a decline in the prey population, and that in turn produces a decline in the predator population. These relationships work so that, over time, predator and prey populations balance each other at stable average levels.

Negative feedback is frequently used in technology to create stable situations, such as a consistently cool temperature in a refrigerator or a consistently warm temperature in a room. Thermostats are technological examples of negative feedback.

A system incorporating many variables regulated by negative feedback is a *well-buffered system*. It can absorb a great many disturbances without becoming unstable. But in natural systems, the capacities of buffers are usually limited. A feedback system consumes materials or energy, and if either one is exhausted, the system may collapse. Water wells are a good example. If we take water from a well, more water pours into it from the surrounding soil. The water level in the well appears to be stable. This can create the false impression that we are dealing with an inexhaustible resource. But at some point the groundwater will run out and the well will not replenish itself. Once this happens, the system may remain broken down for a long time or be permanently destroyed. This is precisely the trap that not only the physicist and the economist in our introduction but many of our Tanaland participants fell into.

The *critical variables* in a system are those that interact mutually with a large number of other variables in the system.[2] They are, then, the key variables: if we alter them, we exert a major influence on the status of the entire system.

*Indicator variables* are those that depend on many other variables in the system but that themselves exert very little influence on the system. They provide important clues that help us assess the overall status of a system.

It is obviously greatly to our advantage to look at a deficiency not in isolation but embedded in its system—helping to drive many positive and negative feedback loops, reflecting changes in underlying critical variables, and sending signals through more visible indicator variables. When we understand the links within a system, we can judge where the roots of certain deficiencies lie and can begin to define our goals more adequately.

We need to know more than just the causal relationships between individual variables in a system, however. We may also need to know what abstractions we can make of the variables—what broad concepts subsume the narrow concepts represented by our variables? It may be useful to know as well which hierarchy of parts and wholes the variable belongs to—of what whole is the variable a part, and of what parts does it in turn consist?

The following example illustrates the importance of this kind of information. A participant in the Greenvale experiment saw a need to increase production in the municipal watch factory. She was a student of literature and knew nothing about manufacturing watches. At first she was at a loss and had no idea how to approach her problem. But then it occurred to her: "Wait a minute. Making watches is a production process. I produce things too. I roll my own cigarettes, for example. How do I do that? I need raw materials: cigarette papers, tobacco, and a little spit. Then I put the raw materials together in a certain sequence—that is, according to a set plan—and to do that I need energy, though not very much to roll a cigarette. But the energy demand for other products can be different. So a production process uses energy and follows a set plan to convert raw materials into a finished product. All right, what are the raw materials needed to make watches? According to what plan are those materials shaped and assembled? What skills do the makers of watches have to have? What kind of energy and how much of it is required?"

By thinking of watch production as analogous to rolling cigarettes, this participant was able to develop a mental picture of watch manufacturing. This gave her a basis for asking further questions about watch production and enabled her to grasp quickly the essentials of the field she had to work in.

This kind of analogous thinking is possible only if we consider things in the abstract. We must understand that making watches is only one narrow form of the broad concept of production process. And what all production processes have in common is using energy to put different materials together according to a set plan.

Thinking by analogy may seem, after the fact, a rather primitive and obvious step, but many of our participants never make use of it and therefore bog down hopelessly in concrete situations. The prerequisite for making connections between watch production and rolling ciga-rettes—and therefore for thinking of useful questions to ask—is an ab-stract understanding of watch manufacturing as a production process. For reflection of this kind it is essential to understand the relationships between broad and narrow concepts, between the abstract and the con-crete. These relationships suggest to us how we can apply knowledge from one field in another.

If we know nothing about wolves except that they are predators and if we know a lot about cats, including the fact that they, too, are preda-tors, then relating both wolves and cats to this broad concept gives us a basis for developing hypotheses about wolves. From this basis we may develop correct hypotheses ("Wolves catch mice") and incorrect ones ("Wolves sit in front of mouse holes for long periods of time without moving a muscle"). But we can always correct false hypotheses. And be-cause false hypotheses can ultimately lead us to correct knowledge, they are better than no hypotheses at all.

Knowledge of the constituent elements of a system can also give us important insights into the structure of that system. Looking at the dif-ferent parts that make up a single element—for example, a fish—can re-veal its relationships with its environment. A fish has a respiratory system; therefore, it needs oxygen, and the oxygen it needs it can find only in the water. A fish has a digestive system; therefore, it passes waste matter. What happens to this waste matter in the water and on the bot-tom of the pond? The answer to that question provides insight into the relationships between the living creatures in a pond and the water, the shoreline, and the bottom of the pond. We can often deduce the rela-tionships between the variables of a system if we establish the parts of one element in the system and consider the relationships of those parts to the environment.

If we consider a system not on the level of its initially obvious elements—fish, water, plants—but on the level of parts that make up those elements, we have moved our investigation up a notch in the degree of detail. In theory, we could go as far as the atomic or subatomic level. Thus we should ask ourselves what the correct degree of detail is for our needs. Should we see a garden pool as consisting of individual creatures, individual goldfish, individual water beetles, individual water lilies? Or should we see it, as in figure 12, as consisting of aggregated animal populations, plant populations, water, and pond bottom? Or should we see it as consisting of the amount of oxygen dissolved in the water, of the nutrients in the water, of the respiratory and digestive organs of the fish? Or should we see it as a "cloud" of elementary particles?

There is no a priori appropriate level of detail. It may happen that in working with a system we will have to move from one level of detail to another. As a rule, however, we should select the level of detail needed to let us understand the interrelationships among our "goal variables," that is, among the variables that we want to influence. We do not need detailed knowledge of the specific components that underlie those variables.

If we want to steer an automobile, for example, we need to know how the position of the steering wheel affects the position of the front wheels. As long as the entire steering system is in proper working order, we don't need to know that the causal connection between the steering wheel and the front wheels is not direct but is in fact made by way of many intermediate parts. For the purpose of steering a car, an understanding of the intricate steering linkage would be superfluous knowledge too detailed for the task at hand. But for a mechanic, an understanding of the different components of a car's steering gear would not be superfluous at all. Even for a mechanic, however, knowledge of the crystalline structure of the steel in the steering column would be too detailed.

To deal effectively with a system:

- We need to know on what other variables the goal variables that we want to influence depend. We need to understand, in other words, how the causal relationships among the variables in a system work together in that system.
- We need to know how the individual components of a system fit into a hierarchy of broad and narrow concepts. This can help us fill in by analogy those parts of a structure unfamiliar to us.
- We need to know the component parts into which the elements of a system can be broken and the larger complexes in which those elements are embedded. We need to know this so that we can propose hypotheses about previously unrecognized interactions between variables.

How do we acquire knowledge about the structure of a system? One important method is analogy, as illustrated above. Another method, and probably the more common one, is to observe the changes that variables undergo over time. If we observe in a given ecosystem that an increase in animal population A is followed by an increase in animal population B and if we then observe a decline in population A followed by a decline in population B, we can assume that animals of type B feed on animals of type A and that the two populations form a predator-prey system. The observation of covariations, between which there may be a time lag, is one way of acquiring structural knowledge, and all it requires is the collection and integrating of data over time.

Even after we know enough about a system to understand its structure, we must continue to gather information. We need to know about the system's present status so as to predict future developments and assess the effects of past actions. These requirements make information essential for planning.

## Solving Problems One
## at a Time

Let's return to the plight of the Moros. Things were not going well for the Moros at the start. They were suffering from many diseases; infant mortality was high; their cattle were undernourished and suffered from a sleeping sickness carried by the tsetse fly; their water supply was very limited. The Moros had to work hard. They had to tend their cattle; plant, hoe, harvest, and grind their millet; gather wood and dry manure for fuel. A difficult and monotonous life.

Using our computer simulation of the Moro territory in the Sahel, we conducted many studies of the Moros' problems. The results pointed more or less conclusively in one direction, and our participants often began by combating the tsetse fly. Their actions led to an increase in the cattle herds. And that led in turn to a still greater shortage of water, which is scarce in the Sahel to begin with, and to more and louder complaints from the Moros about the shortage. At this juncture, at the latest, most participants began drilling wells to alleviate the shortage. (Some participants reversed the sequence. They began by drilling wells; then, when the expected increase in cattle did not materialize, they took measures against the tsetse fly.)

Some of the wells were equipped with pumps to increase their yield. The water shortage eased. The cattle had enough water; the vegetated area increased; the fields could be irrigated better; more millet could be grown. Not only was the Moros' food problem solved, but they had surplus cattle and millet to sell. The sales enabled them to recoup the money invested in the irrigation project and even turn a profit. In short, the whole scheme paid off.

Our participants now felt it was high time to let the Moros reap some benefits from the economic surplus. The first thing most participants thought of was medical care. The Moros received instruction in basic hygiene, and when sufficient funds were available a kind of airborne ambulance service was established to provide quick help in emergencies. A hospital was perhaps more than the Moros needed, but a well-equipped aid station manned by appropriate personnel seemed in order. The consequences were soon evident: the Moros no longer died from minor infections, tetanus, or snakebite. Infant mortality declined rapidly; life expectancy increased. There were further consequences, of course: the population grew markedly. Up to a point, that growth was desirable. The government was happy to see more people living in the sparsely settled border territories but on the whole did not want the population to be too large. It wasn't long before the population explosion made our participants uneasy, and they tried to stem it. But the process was slow, because the Moros were reluctant to give up their old ways and were not easy to convince of the need for birth control. After all, things were going well for them. But despite the Moros' resistance, the birth-control measures had some effect and population growth slowed somewhat. From this point on, the rest of the story can move in a number of possible directions, but almost all of them lead to the same place.

1. *Cattle catastrophe.* The number of cattle continues to grow and ultimately exceeds the carrying capacity of the available vegetated area. Drilling more wells is of no further help because grass will not grow on rocks no matter how much water is poured on them. Now the starving cattle are not only eating the grass but are tearing it up by the roots. The result is massive ecological damage. The vegetated area shrinks at an alarming rate because of the destructive effects of overgrazing. The pattern is a classic example of positive feedback: the smaller the vegetated area becomes, the more desperate the hunger of the remaining cattle and the greater the ecological damage to the sod.

In a situation like this it is a rare participant who will hit upon the correct but initially radical and seemingly "excessive" measure of either slaughtering or selling almost the entire herd to rescue the remaining pasturage. Few participants interpret the trend over time, and unfortunately, so drastic a measure continues to seem excessive and inappropriately brutal until the point is reached where it will no longer do any good.

Once the grass disappears, the cattle do too. The Moros are short of food and have to import it. But because the loss of their cattle also means the loss of one of their two export goods, they are also short of money. Without outside help, famine is a certainty.

Figure 13 shows how the measures taken by the participant with the code name pmost611 resulted in just such a catastrophe. We see here how the number of cattle (black squares) grows significantly without the vegetated area (black inverted triangles) keeping pace. In year 12 the turning point is reached. The cattle and the grasslands, both regulated by positive feedback, disappear. The minimal arable land is far too small to feed the Moros, and in year 18 they experience a famine that wipes out practically the entire population.

It should give us pause to realize that *without* the well-meaning efforts of this participant the Moros could have gone on living in this region indefinitely. Their standard of living might have been lower, but they would not have starved.

A simulation game like this, whose pawns are a fictional starving

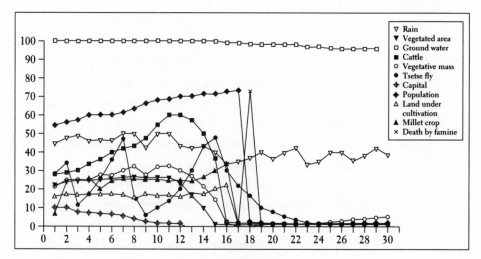

**Fig. 13.** A cattle catastrophe

people, may well seem frivolous, if not macabre. But the situation in the Sahel and in Ethiopia is far more macabre. There, the victims of measures like these are real.

Presented with the abstract mathematical structures that underlie these simulation games, participants in our experiments would not have displayed certain of the behaviors that the semantic trappings of the variables elicited from them. For example, participant pmosc606, whose record is shown in figure 15, refused for ideological reasons to do anything that would harm the ecology of the Moros' territory, and consequently he declined to take any measures against the tsetse fly or to drill any wells. He would have had no such scruples had he been confronted with an abstract mathematical system. When this participant saw that his proecological intentions produced antiecological results, he was obliged to rethink his position. A similar failure in dealing with an abstract mathematical task probably would have moved him to cite his lack of mathematical ability or some other such cause. While he considered himself knowledgeable about ecological matters, he likely had much less pride invested in the notion of himself as a mathematician.

2. *Groundwater catastrophe.* There are, of course, participants who proceed cautiously with the cattle and keep the herds at a level appropriate to the available grasslands. If they can manage at the same time to contain population growth and not let it increase too rapidly—a situation that would prompt the Moros to demand more cattle—then they will maintain stability for a considerable length of time. The newly drilled wells supply enough water for both the cattle and the millet fields. The whole situation looks rosy and appears to be very stable.

At some point, however, the wells start yielding less water. Because it is impossible now either to stop irrigating the millet fields or to stop giving the cattle water to drink, the only solution is to drill more wells. This solution seems all the more obvious because the drop in yield from the existing wells is relatively small and occurs, perhaps, in a year with a little less rainfall than in previous years. Because the indicator is weak, in other words, and because it may well be due to temporary changes in the climate and in water consumption (maybe the Moros had a few more cattle this year), it is easy to interpret the reduced yield as a local variation and to overlook it as an indicator of a failing groundwater supply.

But here, too, the effects of positive feedback are soon felt. Because the groundwater supply is in reality already seriously depleted, the new wells only accelerate its depletion. Soon almost all the groundwater is gone. The wells are practically dry, and the results are comparable to those of the cattle catastrophe. The grasslands shrink, the cattle destroy the sod, and so forth. Catastrophe overtakes the participant rather suddenly because he has either not noticed or not taken seriously the gradual decline of the water supply over a period of years. (The algae infestation of the North Sea in the spring of 1988 seemed to many people to have come out of nowhere too.)

To drill new wells in the face of a declining yield from the existing ones is to create a positive-feedback loop: the less water there is, the more wells we will drill, and the less water there will be. The depletion of the groundwater creates a second positive-feedback loop: the less water and the less vegetation there are, the greater the hunger of the cattle and the more widespread the destruction of the sod. The upshot is that the Moros' cattle—their main source of nourishment—die off and the Moros are left with empty hands and empty stomachs.

Figure 14 shows such a development in its early stages. This participant has created a massive boom. In year 19 the size of the cattle herd goes off the top of the chart. The population shows strong growth, and capital is abundant. About year 22, however, the number of cattle begins to decline. Groundwater has been declining steadily since year 13. The expansion of vegetated land levels off in year 20, and the millet crop is growing too slowly to be able to feed the entire population. Already, we can easily see how this story will end. But the simulation up to this point will leave our experiment participant feeling that he was born to be a development director in a Third World country, for at the conclusion of the experiment in year 30 things are still going swimmingly for the Moros. (We can easily reverse the downturn in the cattle

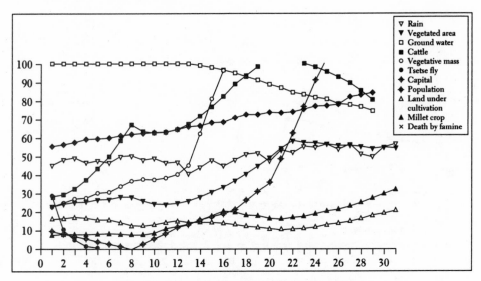

Fig. 14. A groundwater catastrophe in the making

herds, our participant thinks. We'll just sell fewer. We have all the money we need as it is. And it certainly can't hurt to drill a few more irrigation wells and enlarge our vegetated acreage a bit.)

3. *Population catastrophe*. Some experiment participants keep the cattle herds at a constant level and manage the water resources prudently but still find themselves faced with rapid population growth that they cannot simply "turn off" at will when it becomes threatening. The Moros, now far more numerous than before, are clamoring for more food. What to do? Maybe we should slaughter more cattle. Maybe we'll have to forgo exporting meat and millet. Maybe we ought to increase the size of the herds or the acreage given over to millet production. Both measures would require more water, but enlarging the herds is the riskier of the two because it often produces either a groundwater or a cattle catastrophe.

Figure 15 exemplifies an almost "pure" population catastrophe. From the outset this participant decides against increasing the herds and allows the tsetse fly to go about its business unimpeded (though the number of tsetse flies declines along with the declining number of cat-

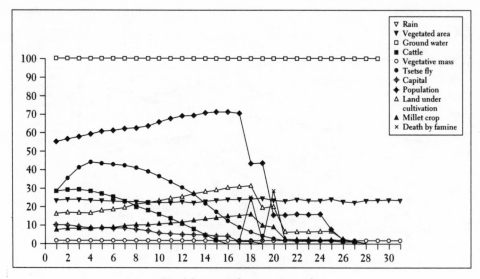

Fig. 15. A population catastrophe

tle). The fact that the vegetated area does not increase shows that this participant has not troubled himself with providing additional irrigation for pastureland. Acreage under cultivation does increase a little, but this increase, which is merely a reflection of population growth, remains insignificant. The only measure this participant enacts energetically is providing the population with good medical care. But pursued in isolation, this move is disastrous. The growing population begins almost immediately to overwhelm its own resources and finally, in years 18, 20, 25, and 26, suffers devastating famines, the end result of well-intentioned measures.

One basic error accounts for all the catastrophes: none of the participants realized that they were dealing with a system in which, though not every element interacted with every other, many elements interacted with many others. They conceived of their task as dealing with a sequence of problems that had to be solved one at a time. They did not take into account the side effects and repercussions of certain measures. They dealt with the entire system, not *as a system* but as a bundle of independent minisystems. And dealing with systems this way breeds trouble: if we do not concern ourselves with the problems we do not have, we soon have them.

If we need more export goods to improve the Moros' financial situation, we have to raise more cattle. If we want more cattle, we have to enlarge our pasturage, and to do that we have to drill wells. Although the problem is solved, the solution simply creates more problems.

If we drill wells in order to provide water to farmers and cattle herders, we encourage a level of millet and cattle production that cannot be sustained with any lesser water supply. Any decrease in the yield of the wells must be offset by drilling more wells if food production is not to suffer. This drilling leads to further decreased yields and consequently to further drilling. Soon the almost nonexistent water supply will support neither cattle nor crops.

If we improve medical care, the population will grow, a change that is, within limits, desirable. Then, too, this improvement simply makes life better for the Moros. But a larger population will in the long run require more food and consequently affect cattle herds, pasturage, and groundwater.

Most participants lack an overall view of the system and of the reciprocal interactions within it. Our Moro game is, however, transparent enough that they realize they have made mistakes the moment they become aware of the catastrophes they have brought about. The game is therefore a useful pedagogical tool. Everyone realizes at a second glance that the water in the wells must come from somewhere and that the reservoir the wells draw on must be refilled if the status quo is to be maintained.

Why do our participants tend to see a bundle of many independent minisystems instead of one overarching system? One reason they deal with partial problems in isolation is their preoccupation with the immediate goals described in the previous chapter. The shortage of water for the cattle is the problem we *have* and the problem we need to solve. At the moment, we don't have other problems, so why think about them? Or, to put it better still, why think that we should think about them?

Another reason is informational overload. Participants are given a lot of information, and to solve their problems, they have to gather a lot of data and address many aspects of the situation. There just doesn't seem to be time enough to worry about problems that are not immediately pressing.

## "It's the Environment"

To deal with a system as if it were a bundle of unrelated individual systems is, on the one hand, the method that saves the most cognitive energy. On the other hand, it is the method that guarantees neglect of side effects and repercussions and therefore guarantees failure. If we have no idea how the variables in a system influence one another, we cannot take these influences into account. The previous section makes this clear. It is obviously better if we know how the individual variables in a system relate to one another.

Figure 16 presents in graphic form a hypothesis about a set of such relationships. A participant in the Greenvale experiment produced this diagram, which, as we can see, takes account of many important variables in the Greenvale system: the productivity of the watch factory, the factory's income, unemployment, the satisfaction of Greenvale's residents, the maintenance of public buildings and grounds, the achievement level of children in school, the amount of help parents give children with their homework, the prevalence of illness in Greenvale . . .

The interesting thing about the diagram is its form. The entire network of interactions can be traced back to a single point, the "satisfaction" of Greenvale's citizens. This factor influences all the others and is the only variable in the system given a central position. If the citizens are content, they will not often fall ill. That means they will work a lot.

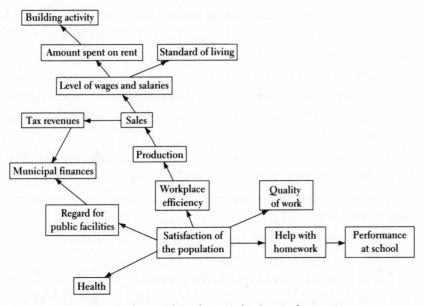

Fig. 16. A "reductive" hypothesis in the Greenvale experiment

And that means the quality of their work will be good and their production high. The watch factory will be able to sell many watches, and that means it will take in a lot of money. It can thus create more jobs. And that means unemployment will disappear. It also means that the factory can pay higher wages. And that means the employees of the factory can afford better apartments, and so the number of building starts rises.

The benefits go on and on. Greenvale's contented parents are willing to give their children considerable help with their homework. The children therefore perform better in school. And therefore the workforce becomes more skilled and the quality of their products improves. Contented citizens are kind to public facilities and feel no need to demonstrate their skills by smashing park benches. This means the public coffers are spared outlays for repairs.

Taken singly, all these connections may be correct, but given its centralistic organization, the hypothesis as a whole is wrongheaded and dangerous. A reductive hypothesis of this kind, tying everything to one variable, has, of course, the positive virtue of being a holistic hypothesis, which is desirable because it encompasses the entire system. But it does so in a certain way, namely, by reducing the investment of cognitive energy. What do we need to focus on in a system organized like the one in figure 16? On one thing and one thing only: the satisfaction of the citizens. The solution of all other problems will follow automatically.

In addition to encompassing the entire system, a reductive hypothesis has the virtue of making it easy to deal with the system. And in its individual assessments this hypothesis is not wrong. But in its overall assessment it is wrong because it is incomplete. It does not take into account the manifold feedback loops in the Greenvale system or the fact that in this system, as in many others, we are dealing not with a star-shaped network of interdependencies but rather with a network that more closely resembles an innerspring mattress. If we pull on one spring, we will move all the others, some more, some less. And if we press down on another, the same thing happens. There is no single central point. Not every point is a central one, but many are.

A second and more careful look should make this perfectly obvious. If we say, for example, that the satisfaction of parents depends on how their children are doing in school and if the children's satisfaction also greatly depends on it, then what we have in the interlocking relationship of "satisfaction," "amount of help with homework," and "performance in school" is a positive-feedback loop. It is also true that the potential for negative-feedback loops exists. The satisfaction of the citizens could make Greenvale more attractive as a place to live and work. That could draw more people to Greenvale from the surrounding area and from other towns, imposing additional burdens on public institutions and services and draining municipal finances. And that could have a negative impact on the quality of life and thus on the satisfaction of Greenvale's citizens.

The point is that the satisfaction of the citizenry is in reality embedded in a network of positive- and negative-feedback loops, and knowing what will maintain citizen satisfaction at a high level in the long run is no simple matter.

The reductive hypothesis shown in figure 16 allows complicated considerations to be avoided. And that is the reason why reductive hypotheses are so popular both in Greenvale and beyond. One reductive hypothesis that produced far more hideous results than any failings in Greenvale ever could was the popularly advanced idea that assorted subversives like Jews, Jesuits, and Freemasons undermined the German army and thereby single-handedly brought about Germany's defeat in World War I. Other examples can be found in abundance. In the 1950s, every exceptionally heavy summer hailstorm was blamed on atomic-bomb tests, which were conducted more frequently then than they are now. If seals die in the North Sea, the ecologically dismal state of the North Sea must be the cause.[3]

In a television play I saw a long time ago and have otherwise completely forgotten, the hero is going home on a streetcar one evening when the car goes around a curve and throws a somewhat tipsy gentleman standing across the aisle off balance. Stumbling into the hero, he apologizes and then blurts out: "It's the environment," an explanation he continues to repeat and summons once again when he realizes he has missed his stop. The world of politics, too, becomes amazingly lucid and transparent once our eyes are opened and we understand that the KGB or the CIA or the communists or the capitalists or whoever are pulling all the strings. The person who feels moved for one reason or another to study the nature of our world or at least of our society and who concludes that we live in an "automobile society" or a "service society" or an "information society" or an "atomic society" or a "leisure society" proffers a reductive hypothesis that invites us to extrapolate a structure from it.

The fact that reductive hypotheses provide simplistic explanations for what goes on in the world accounts not only for their popularity but also for their persistence. Once we know what the glue is that really holds the world together, we are reluctant to abandon that knowledge and fall back on an unsurveyable system made up of interacting variables linked together in no immediately obvious hierarchy. Unsurveyability produces uncertainty; uncertainty produces fear. That is probably one reason people cling to reductive hypotheses. People use many dodges to defend their pet hypotheses against logical argument or the evidence of experience.

One excellent way to maintain a hypothesis indefinitely is to ignore information that does not conform to it. Once our satisfaction-oriented participant (see fig. 16) had proposed his hypothesis, he proceeded to study the Greenvale watch factory and to discover the reason for its low productivity. In keeping with his central hypothesis, his inquiry consisted of asking the blue- and white-collar workers in the various divisions of the plant how satisfied they were. When one worker ("simulated" by the director of the experiment) complained about the poor condition of the plant's machinery, our participant replied, "Yes, yes, but are your coworkers as dissatisfied as you are?" This mayor never came back to the topic of inadequate machinery, which was in fact the reason for the factory's low productivity.

We are infatuated with the hypotheses we propose because we assume they give us power over things. We therefore avoid exposing them to the harsh light of real experience, and we prefer to gather only information that supports our hypotheses.[4] In extreme cases, we may devise elaborate and dogmatic defenses to protect hypotheses that in no way reflect reality.

## Prime Numbers and Tourist Traffic, or Moltke and Forest Fires

In 1640 the French jurist and amateur mathematician Pierre de Fermat, who was in fact one of the most important mathematicians of the seventeenth century, wrote a letter informing his colleague Marin Mersenne that he had developed a procedure for finding prime numbers (that is, numbers that can be divided evenly only by themselves and by 1). For each of the numbers 0, 1, 2, 3 . . . , Fermat produced a "Fermat number" $F_0$, $F_1$, $F_2$, $F_3$. . . . The method for producing these numbers can be summarized by the following formula:

$$F_n = 2^{2^n} + 1,$$

where n ranges over the numbers 0, 1, 2, 3. . . . So, for example, $F_0 = 2^{2^0}$ $+ 1 = 2^1 + 1 = 2 + 1 = 3$, $F_1 = 5$, $F_2 = 17$, and $F_3 = 257$. Note that the numbers are increasing; this trend continues, and as n grows, the Fermat numbers become enormous. The reader can verify, as Fermat undoubtedly did, that these first four Fermat numbers are indeed prime. If Fermat had carried his calculations further, however, he would soon have seen that not all the numbers generated by his formula are prime. For $n = 5$, $F_5 = 4,294,967,297$, which is not a prime number, as the Swiss mathematician Leonhard Euler was able to demonstrate in 1732. (With numbers of this magnitude, the arithmetic is difficult, of course.)

Fermat made the mistake of overgeneralizing here, a very common error in formulating hypotheses. We find example 1, which has certain features. Then we find example 2, which has the same features. And then we find examples 3 and 4, which also have the same features. And so we conclude that every conceivable example of this type will have the same features.

The formation of abstract concepts by means of generalization is an essential mental activity. We could not begin to cope with the multitude of different phenomena we encounter if we did not put them together in categories. If we had to determine whether an object before us was a chair every time we were faced with an object that looked like other objects we had been given to understand were chairs, we would not get very far in our daily endeavors. We need an abstract concept "chair" that lets us dispense with complex deliberations and simply operate with the idea of a chair, whether we have a concrete example in front of us or not.

The ability to identify common characteristics in only a few examples of a certain type of thing and then to formulate an abstract concept on that basis is very useful, and without this ability we would be overwhelmed by the variety of phenomena we encounter. We need an abstract concept "chair" that lets us ignore the color of the slipcover, the fabric of the upholstery, what the legs are made of, and so forth, and judge the "chair-ness" of an object only on the basis of whether it has four legs, a surface to sit on, and a backrest, all in the proper proportions and relationships.

The world of our ideas is usually not particularly colorful. Our ideas are pale outlines, and even the mental image of something as vivid as a rose cannot compare in intensity with the perception of a real rose's color and contours. We may regret this but it has its advantages. These hollow, schematic constructs represent classes of equivalence. Their schematic character allows us to regard very different roses (or lamps or pencils or teacups) as more or less the same.

A few examples often suffice to provide us with an abstract grasp of something. Four examples were enough to convince Fermat he had discovered a *general* procedure for generating prime numbers.

Essential as it is to put aside "unimportant" features and to stress "important" ones in formulating classes, the dangers of this intellectual operation are great. A necessary generalization can easily evolve into an overgeneralization. And as a rule we have no opportunity to test in advance whether a concept we have developed has struck just the right degree of abstraction or is an overgeneralization.

One participant in the Greenvale experiment had achieved favorable results with the promotion of tourism. The construction of a few hotels and encouragement of bed-and-breakfast establishments had attracted tourists and brought considerable overall improvement in the town's financial status.

The experience of success left an indelible mark. "The promotion of tourism," our participant believed, "pays off." But this abstract formulation was an overgeneralization based on only one success story.

His promotion of tourism had proved successful only because it had coincided with a favorable constellation in the environment: there had been people in Greenvale at the time who could devote their energies to the tourist business, and there had also been a demand for that business outside Greenvale. But our participant had not taken note of that constellation of conditions in formulating his abstract concept. All he had registered was the success of a general "if . . . then" rule: "If I promote tourism, then sooner or later I'll have more money in the town coffers."

This "deconditionalized" concept—this concept removed from the context of conditions bearing on it—led our participant into disaster. When some later missteps on his part brought Greenvale to the brink of bankruptcy, he invested all available money in a massive tourism campaign. But because the conditions that had made tourism profitable before were no longer present, the money yielded no appreciable return. The English psychologist James T. Reason thinks that this kind of error is the result of a general propensity for "similarity matching," that is, a tendency to respond to similarities more than to differences.

Our participant's behavior is perhaps an extreme case but not an unrealistic one; indeed, it is altogether understandable.

In complex systems with many interlocking elements, deconditionalizing abstractions are dangerous. The effectiveness of a measure almost always depends on the context within which the measure is pursued. A measure that produces good effects in one situation may do damage in another, and contextual dependencies mean that there are few general rules (rules that remain valid regardless of conditions surrounding them) that we can use to guide our actions. Every situation has to be considered afresh.

A good example of a system in which context should determine strategy is the "fire" simulation game, which we modeled on a similar game developed by Bernd Brehmer of the University of Uppsala. In this game the "fire chief" has to manage his twelve brigades so that as much of the woods as possible and, more important, the village are protected from forest fires. It is a dry summer, and fires can break out anywhere at any time. With the help of satellites, the chief can survey the entire district and issue commands to his units by radio. He can, for example, order his brigades to travel to a certain point, to keep an independent lookout for fires (within their limited field of vision), to fill their water tanks, to patrol a certain area, to put out a fire on their own or to pass through a fire line without putting the fire out, and so on. The chief can also obtain information at any time about the present status of his units (amount of water on hand, current mission, etc.).

The main characteristic of the chief's situation in this game is that most of the measures he can take are effective only in particular contexts. In one set of circumstances, a certain measure may be appropriate. In different circumstances, perhaps just the opposite measure is called for. Sometimes it is advantageous to bring all the units together and keep them together. Sometimes it is advantageous to spread them out.

It is a good idea to distribute the units widely if we want to have them in position to put small fires out quickly before they can spread. This strategy makes sense if there is no wind to spread a fire and if the entire area to be protected is small relative to the number of brigades available. On the other hand, it makes sense to concentrate the brigades in one place if the entire district cannot be covered anyway or if the wind is so strong that any fire will spread rapidly. In the latter case, many units will have to be concentrated quickly at a fire's point of origin to battle it effectively.

Sometimes it is advantageous to combat a spreading fire head-on. This is the proper strategy if we have enough units and enough water available and if the fire is not too large or the wind too strong. At other times, however, it may be better to attack a fire on its flank and let it run in its major direction of spread. This would be appropriate if both the fire and the wind were strong and we had only a few units available to fight it. With a flanking action, we can at least guide the fire to some extent; a frontal attack under unfavorable circumstances would be both futile and dangerous.

If we have to fight two or more fires at once, it may sometimes be advantageous to dispatch individual units to fight them one on one, assuming we have enough units and enough water and the units are close enough to these relatively small fires to reach them quickly. It may sometimes be better, however, to let one fire go so that we can concentrate our resources on another. In this case, we would not spread our units out but would bring them together to fight the one fire. This is the proper strategy if we are short of units, if the distance to another fire is great, if we are low on water, and if, perhaps, the wind is driving one of the fires onto barren ground where it will go out by itself anyway.

A fire chief thus encounters situations in which one strategy is called for and others in which just the opposite is called for. A wise decision will always depend on the given constellation of conditions, and to reach that decision a fire chief must consider the present location of his units, the wind direction, the size of the fire, the amount of water the units have with them, the speed at which the different units can travel, and the distances they have to cover. Experiment participants who try to use general, deconditionalized measures in a system like this will fail in the long run. A rule such as "Brigades should at all times be widely distributed over the district" is too general to be useful, and measures based on it will be wrong much of the time. The rules for action that apply here have to be more of the type "If A and B and C and D are the case, then X. But if A and B and C and E are the case, then Y. And if A and F and C and D and E, then Z."

Situations of this kind are doubtless what the nineteenth-century Prussian field marshal Graf von Moltke had in mind when he wrote, "Strategy is a system of makeshifts. It is more than a science. It is bringing knowledge to bear on practical life, the further elaboration of an original guiding idea under constantly changing circumstances. It is the art of acting under the pressure of the most demanding conditions. . . . That is why general principles, rules derived from them, and systems based on these rules cannot possibly have any value for strategy."[5]

What Moltke had in mind about strategic thinking in war applies in general to the manipulation of highly interdependent systems. Schematizations and the formulation of rules obscure the constant need to adapt action to context. A sensible and effective measure in one set of circumstances can become a dangerous course of action when conditions change. We must keep track of constantly changing conditions and never treat any image we form of a situation as permanent. Everything is in flux, and we must adapt accordingly. The need to adapt to particular circumstances, however, runs counter to our tendency to generalize and form abstract plans of action. We have here an example of how an important element of human intellectual activity can be both useful and harmful. Abstract concepts are useful in organizing and mastering complicated situations. Unfortunately, this advantage tempts us to use generalization and abstraction too freely. Before we make a generalization, we should consider whether we have enough evidence to do so. Before we apply an abstract concept to a concrete situation, we should submit it to "strategic" scrutiny to decide whether it is appropriate to the context.

## The Pale Cast of Thought

> And thus the native hue of resolution
> Is sicklied o'er with the pale cast of thought,
> And enterprises of great pith and moment
> With this regard their currents turn awry,
> And lose the name of action.
>
> Hamlet, III, i

Anyone who has a lot of information, thinks a lot, and by thinking increases his understanding of a situation will have not less but more trouble coming to a clear decision. To the ignorant, the world looks simple. If we pretty much dispense with gathering information, it is easy for us to form a clear picture of reality and come to clear decisions based on that picture.

Sometimes there is probably even positive feedback between the amount of information we have and our uncertainty. If we know nothing at all about something, we can form a simple picture of it and function on that basis. Once we gather a little information, however, we run into trouble. We realize how much we still don't know, and we feel a strong desire to learn more. And so we gather more information only to become more acutely aware of how little we know. . . .

The self-reinforcing feeling of uncertainty and insecurity that results probably accounts for many an unfinished dissertation or book. As we gather more and more information, our conviction that we have formed an accurate picture of the world gradually gives way to doubt and uncertainty. Is acid rain really the cause of forest mortality? What produces acid rain? Only auto emissions? If not, what else? How do the root systems of trees actually work? How exactly do plants and trees absorb nutrients?

The more we know, the more clearly we realize what we don't know. This probably explains why we find so few scientists and scholars among politicians. It probably also explains why organizations tend to institutionalize the separation of their information-gathering and decision-making branches. A business executive has an office manager; presidents have councils of advisers; military commanders have chiefs of staff. The point of this separation may well be to provide decision makers with only the bare outlines of all the available information so that they will not be hobbled by excessive detail when they are obliged to render decisions. Anyone who is fully informed will see much more than the bare outlines and will therefore find it extremely difficult to reach a clear decision.

This positive feedback between information gathering and uncertainty does not occur if we are able to learn everything, or at least everything important, about a sector of reality in the time available to us. We are fully informed then and can spin out all the possible consequences of any decisions we might make. But in reality we rarely have complete information. We don't know the status of certain variables because they are invisible to us. How can we quickly gather precise information about the groundwater supply in a given region? How can we find out how many automobile owners will set out for what vacation destinations in 1997? How can we know what our opponents intend to do next if they are careful to keep their intentions secret?

Positive feedback between uncertainty and information gathering may explain why people sometimes deliberately refuse to take in information. It is said that before the Seven Years' War Frederick the Great declined to hear about the modernization of Austrian and Russian artillery.[6] And it is said that before his invasion of Poland Hitler deliberately ignored a report that England was serious about coming to the aid of its ally if Germany attacked Poland.

New information muddies the picture. Once we finally reach a decision we are relieved to have the uncertainty of decision making behind us. And now somebody turns up and tells us things that call the wisdom of that decision into question again. So we prefer not to listen.

The results of an experiment devised by Rüdiger von der Weth reveal very clearly this inverse relationship between gathering information and readiness to reach decisions.[7] The participants in von der Weth's experiment had to learn, in a limited amount of time, how to operate a complicated machine that processed raw "lithum" into salable products. The machine was fueled by "lithanol" pumped to the motors of four separate processors, one for each stage of production. The stages were represented by the four processed forms that the lithum took as it passed through the machine: lithum powder, lithum lye, lithum tar, and lithanol (which could be recycled into the machine's fuel system). Each of the four forms could be sold, and at each stage of production the participant had the option of drawing off some quantity of lithum product for sale while allowing the remainder to continue through further processing.

By adjusting the speed of the motors and the settings of the processors and the draw-off valves, the experiment participants were to control production so as to yield the maximum possible profit. To accomplish this, the participants could adjust the parameters of the individual processors or shut one or more processors down altogether. (No prior technical training was required to operate this simulated machinery.)

The exhaust system played a special role. All the emissions from the motors were collected in a container and could be detoxified by catalyzation. This process was relatively expensive and reduced the machine's profits. The exhaust gases could also be released into the atmosphere without first undergoing detoxification.

The participants' task was to set the parameters of the machine to yield the highest possible profits while imposing the smallest possible burden on the environment. These are, of course, contradictory goals. In addition, our participants had to acquire, in a limited amount of time, a vast amount of information about not only technical but also economic issues.

The graph on the left in figure 17 shows that the bad participants made significantly more decisions in the first five phases of this experiment than did the good participants, and the graph on the right shows that the bad participants asked significantly fewer questions than did the good ones. In short, the bad participants displayed, at least in the early phases of the experiment, a reluctance to gather information and an eagerness to act. By contrast, the good participants were initially cautious about acting and tried to secure a solid base of information. What we plainly see here, then, is an inverse relationship between information gathering and readiness to act. The less information gathered, the greater the readiness to act. And vice versa.

Behavior of this kind has its consequences. Figure 18 records the degree of attention the good and bad participants gave to production and

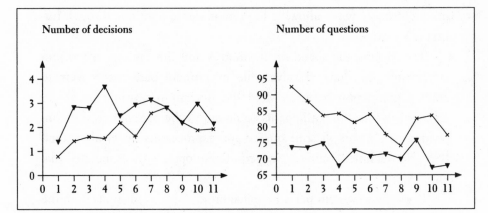

**Fig. 17.** Questions and decisions of good (+−+−+) and bad (▼−▼−▼) participants
in von der Weth's experiment

to emissions detoxification over the course of the experiment. It is obvious from the graphs that the good participants displayed both a clear fo-  ·
cus and a clear shift in focus. They concentrated on production at first
and then turned, in the second phase of the experiment, to minimizing
the environmental impact. No clear shift in focus is evident with the

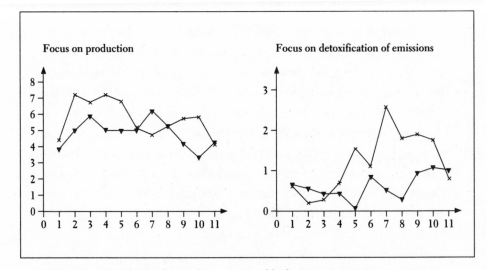

**Fig. 18.** Areas of focus for good (+−+−+) and bad (▼−▼−▼) participants over time

bad participants, and throughout the experiment they gave considerably less attention to the environmental problem than did the good participants.

The relationship between success, level of information gathering, and readiness to act can also be the opposite of that in the lithum experiment, however. Figure 19 shows the number of decisions and the number of questions over eight phases of the Greenvale experiment. In this case, the good participants made more decisions and asked fewer questions than the bad ones. That is, they behaved like the bad participants in the lithum experiment.

Why? In my view, the difference lies in the time constraint, which was much more pressing in the lithum experiment. In both experiments, the good participants gathered *enough* information to let them make necessary decisions. The bad participants reacted to the time pressure of the lithum experiment by refusing to gather information and by leaping into action. In the Greenvale experiment, however, bad participants responded to the absence of time pressure by gathering too much

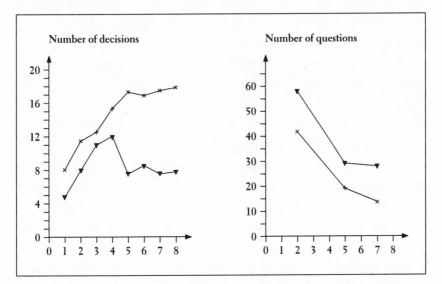

Fig. 19. Questions and decisions of good (+−+−+) and bad (▼−▼−▼) participants in the Greenvale experiment

information. The excess information bred uncertainty, the uncertainty moved them to gather still more information, and that information inhibited their decision making all the more. The fact that the number of decisions the bad participants made after the fourth session sank rather than rose speaks for this interpretation.

The two modes of behavior are opposite sides of the same coin. We combat our uncertainty either by acting hastily on the basis of minimal information or by gathering excessive information, which inhibits action and may even increase our uncertainty. Which of these patterns we follow depends on time pressure or the lack of it.

The carousel of positive feedback, of information gathering and increasing uncertainty, does not spin eternally. If we are not able to satisfy ourselves at some point that we do in fact have enough information, we finally throw in the towel. We simply stop playing a game that makes us more and more uncertain and anxious and less and less capable of action. We may resign ourselves to total inaction or we may give in to irrationality and base our actions on intuition. Enough rationality—all this information gathering is useless! Listen to what your heart is telling you and let your intuition be your guide! (Which means that you won't know exactly what your guide is.)

We may resort to "horizontal flight," pulling back into a small, cozy corner of reality where we feel at home, like the Greenvale mayor trained in social work who finally focused all her attention on one troubled child. Or we may resort to "vertical flight," kicking ourselves free of recalcitrant reality altogether and constructing a more cooperative image of that reality. Operating solely within our own minds, we no longer have to deal with reality but only with what we happen to think about it. We are free to concoct plans and strategies any way we like. The only thing we have to avoid at all costs is reestablishing contact with reality.

Anyone interested in studying a case in which resignation-induced torpor, refusal to gather and analyze information, distortion of information, and sudden, frantic fits of action all occur simultaneously should read the sections of Joseph Goebbels's diary written during the last days of the Third Reich.[8] There we find "vertical flight" into fantasies about Frederick the Great, Czarina Elizabeth, and the final phase of the Seven Years' War. We also find "horizontal flight": concern about trivial everyday matters no longer of any importance and about the designing of new military medals. And we find, finally, the enactment of utterly brutal military measures that, quite apart from moral considerations, were utterly pointless.

# Five

## Time Sequences

### Time and Space

We live and act in a four-dimensional system. In addition to the three dimensions of space, this system includes the fourth dimension of time, which moves in one direction, and that direction is toward the future. We recognize forms in both time and space. We live with chairs, tables, houses, cars, streets, and trees, configurations in space that remain relatively unchanged over time. A melody, on the other hand, is an example of a configuration in time. It extends over time, and the relationship between tones in a sequence is what constitutes its special character. We rarely have trouble dealing with configurations in space. If we're not entirely sure of what we're looking at, we can take another look and resolve our uncertainty. We can normally look at forms in space again and again and in this way precisely determine their particular configuration. That is not true of configurations in time. A time configuration is available for examination only in retrospect.

But life forces us to try to understand patterns in time. Businesses must pay attention to trends in sales, markets, and production. Any kind of strategic planning must take account of developments over time. Meteorologists, seismologists, demographers, politicians, insurers and the insured, and anyone saving up to build or buy a house — the list of people functioning more or less successfully as prognosticators or even prophets is long. Their challenge is to recognize time configurations as they unfold.

It's late afternoon. I'm sitting in my office, drinking a cup of coffee, putting my feet up, and looking forward to a quiet evening at home. I'm reviewing the events of the day and reflecting on a faculty meeting in which there was a sharp exchange of words on the subject of assigning classrooms. Colleague A attacked colleague B's ideas rather forcefully. Colleague C, a friend of B's, responded sharply and was quite rude to A . . . and then. . . . Suddenly the "logic" of this exchange became clear to me. A had attacked B, and in response C had insulted A. The insult made a bad impression on the others present. Because of this, C reduced his chances of winning approval for his own poorly argued motion later in the meeting. He needed the goodwill of those present to win passage, and this is precisely what he had squandered. C's motion, however, was in conflict with A's interests. So perhaps what I had witnessed was a subtly devised strategy on A's part. He had exploited C's tendency to overreact.

Only now, as I looked back on the meeting, could I impose some kind of structure on the sequence of events: colleague A had made use of the tried and true debating strategy "Make your opponent mad; then maybe he'll make a mistake." Time configurations develop, obviously, over time. When they are only half completed, we cannot predict with certainty what their final form will be. It is also difficult in the thick of developing events to leap back and forth in time, now looking into the future to speculate on what is likely to happen, now looking into the past to review what has already happened. That space configurations can be perceived in their entirety while time configurations cannot may well explain why we are far more able to recognize, and deal with, arrangements in space than in time.

Because we are constantly presented with whole spatial configurations, we readily think in such terms. We know, for example, that to determine whether a parking lot is crowded we need to look at more than one or two spaces. Our experience with spatial forms also gives us great intuition about "missing pieces." If shown an incomplete spatial pattern, we will usually be able to identify it as incomplete and will often have ideas about how to complete it based on notions of symmetry (and asymmetry), repetition, and the like.

By contrast, we often overlook time configurations and treat successive steps in a temporal development as individual events. For example, as enrollment rises each year, the members of a school board may add first one room, then another onto the existing schoolhouse because they fail to see the development in time that will make an additional schoolhouse necessary. Even when we think in terms of time configurations, our intuition is very limited. In particular, our ability to guess at missing pieces (in this case, future developments) is much less than for space configurations. In contrast to the rich set of spatial concepts we can use to understand patterns in space, we seem to rely on only a few mechanisms of prognostication to gain insight into the future.

The primary such mechanism is *extrapolation from the moment*. In other words, those aspects of the present that anger, worry, or delight us the most will play a key role in our predictions of the future. For example, the oil shortage of 1979 prompted the biochemistry professor and science-fiction writer Isaac Asimov to predict that in 1985 the world's demand for oil would exceed the world's production. (In fact, the trend in the price of gasoline, adjusted for inflation, has been downward since the early 1980s, implying a healthy excess of supply over demand.)

Two factors come together in extrapolations from the moment: first, limited focus on a notable feature of the present and, second, extension of the perceived trend in a more or less linear and "monotone" fashion (that is, without allowing for any change in direction). Fixation on the characteristics of the moment brings with it the danger that too much significance is ascribed to present circumstances. A tourist in Hong Kong during the typhoon season, for example, could well become convinced of the colony's imminent watery end; any resident, however, would view the heavy rains as unremarkable in the context of an entire year's weather. Fixation on linear future development may prevent us from anticipating changes in direction and pace: a purely linear extrapolation of the growth of a six-year-old child would produce ridiculous predictions about her height as, say, a forty-year-old.

Our ultimate concern in this chapter is how people form their ideas of the future. If we can identify the typical difficulties people have in dealing with time and in recognizing temporal patterns, we can suggest ways to overcome these difficulties and to improve temporal intuition.

## Lily Pads, Grains of Rice, and AIDS

Children, and many an adult, will be amazed at the answer to the following problem. There is one water lily growing in a pond with a surface area of 130,000 square feet. In early spring, this lily has one pad, and each lily pad has a surface area of 1 square foot. After a week, the lily has two pads; after the following week, four pads. After sixteen weeks the pond is half covered. How much longer will it take before the whole pond is covered?

If we assume that the water lily will continue to spread at a constant rate, then the pond will be covered in only one more week because up to this point the surface area of the lily pads has doubled each week. Obvious as this may seem, the problem still stumps many people. If it has taken the water lily sixteen weeks to cover half the pond, they reason, then how in the world can it manage to cover the other half in only one week?

And then there is the story about the inventor of chess and his master, who was an Indian king. After the inventor had presented the game, the king condescendingly promised the man a reward. The good fellow could select any item he liked from the king's treasury.

The inventor was annoyed at the king's patronizing reception of his achievement, and he devised a subtle revenge. He asked for a very modest reward. No gold, no jewels, no fat sinecure. All he asked was a little rice—one grain for the first square on the chessboard, two for the second, four for the third, eight for the fourth, and so on for all the squares on the board.

The king, delighted to get off so cheap and laughing up his sleeve at the stupidity of the inventor, called for a bowl of rice. It soon became obvious that the bowl contained far from enough rice. And a few calculations made by the court mathematician soon revealed that the inventor's "very modest" request could not be met. For the last square alone, $2^{63}$ grains of rice would be needed, that is, about 9,223,372,036,000,000,000 grains. That amounts to about 153 billion tons of rice if we assume that about sixty grains weigh one gram. Or seen in still other terms, it amounts to about 31 million cargo ships full if each ship holds 5,000 tons. And that is only the amount of rice for the last square on the chessboard. The next-to-the-last square, of course, would take much less, only about 4,611,686,018,000,000,000 grains, a mere half of what would be required for the last square. To call the reward the inventor of chess had asked for "royal" would be, to put it mildly, an understatement.

The shrewdness of the inventor, however, is not as interesting as the naïveté of the king. He was clearly not able to recognize the features of a certain kind of development—exponential growth. A quantity is said to be growing "exponentially" when its value at any time (or for any square of the chessboard) is its previous value multiplied by a particular number, the same number each time. For both the lily pads and the grains of rice, the number was 2—the area of the pads and the number of grains doubled at each stage. (Note that exponential growth is very different from linear growth. In a linear process, a quantity increases by the same *amount*, not the same *multiple*, at each step. Suppose, for example, a ten-pound cat suddenly begins to grow and gains a pound each year. At the end of the first, second, and third years of this process, he weighs eleven, twelve, and thirteen pounds; the *amount* of his increase has been constant. The *multiples* by which he has increased over the same three periods, however, have been $11/10 \sim 1.1$, $12/11 \sim 1.091$, $13/12 \sim 1.083$. If the cat continues to grow, this multiple will approach—though never reach—1, as the single pound gained becomes negligible compared to the great weight of the cat.)

The examples given of exponential development may seem fanciful, and perhaps it is understandable that people have difficulty predicting the outcomes of the amazing lily's growth and the subtle inventor's reward. Even when exponential growth takes a far more mundane form, however, it seems to retain its mystery.

An alternative way of measuring exponential growth is to express the amount by which the quantity increases over its previous value as a percentage of that previous value; this percentage is called the *rate of growth*. For example, both the lily pads and the grains of rice have rates of growth of 100 percent. "Rate of growth" brings us to the familiar vocabulary of interest rates, inflation rates, and the like. The amount of rice needed for the last square of the chessboard can now be characterized as the return on an initial investment of one grain paying 100 percent interest compounded over sixty-three periods (the first square corresponds to the initial investment, the second square to the first period, and finally the sixty-fourth square to the sixty-third period); this quantity can now be calculated according to the standard compound-interest formula.[1]

Unfortunately, we have reason to believe that switching to the language of rates of growth, investments, and yields does not make exponential growth more tractable. A number of psychological experiments have demonstrated that an incapacity to deal with nonlinear time configurations is a general phenomenon. In one of our experiments, we gave the participants the task of estimating the yield of a 6 percent growth rate over a hundred years. The task was presented with the following instructions:

> The management of a small tractor factory believes that it will have to increase its production by 6 percent annually if it is to ensure its long-term existence. In 1976, it produced 1,000 tractors. Please estimate, without making complex calculations, how many tractors the factory will have to produce in the years 1990, 2020, 2050, and 2080 in order to maintain this growth rate.

Figure 20 shows the average results of this experiment in estimation: the participants grossly underestimated the growth that would actually be required. We can conclude from these results that the average newspaper reader, for example, does not understand the information given when an article reports that the U.S. Federal Reserve expects the American economy to grow at the rate of 2.5 percent per year. He thinks he understands, but he doesn't.

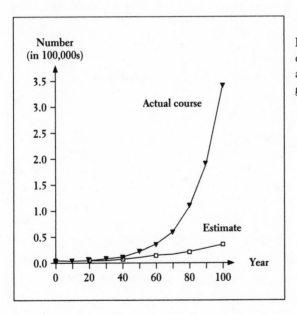

**Fig. 20.** Average estimate of growth at 6 percent and actual course of such growth

People, it seems, even find estimating past developments difficult. Psychologist Andrea Bürkle gave experiment participants figures on oil extraction at the beginning of this century, that is, in the early days of motorization.[2] She also told them that the growth of oil production had remained constantly exponential since that time. In other words, it had grown according to the compound-interest formula, and Bürkle asked her academically trained participants if they understood the formula. They said they did. Bürkle then asked them to estimate the growth of oil production since the beginning of this century. Figure 21 shows the results, once again a gross underestimate of what actually happened.

Lily pads, grains of rice, imaginary factories, and past oil prices may not be very important, but AIDS is, not only for those stricken by the illness but also for the communities and governments that need to supply the funds for research and for care of the sick. Here too, however, we find unawareness of the explosive course exponential development can take.

In 1985, the author of a sober and well-researched newspaper article

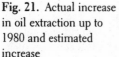

Fig. 21. Actual increase in oil extraction up to 1980 and estimated increase

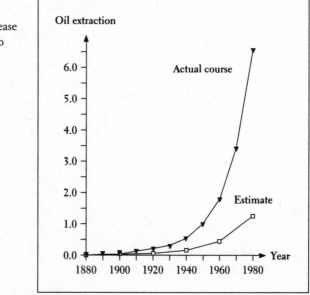

on the spread of AIDS wrote that by 2 September of that year 262 cases of AIDS had been reported in West Germany.[3] In mid-August, the figure had been 230. Of those who had fallen ill, 109 had died. The author ended her article by asking if that wasn't a small number compared with the number of deaths by cancer, traffic accidents, heart disease. Here again we find the condescending smile of the Indian king: who can get worked up over such a little bit of rice?

Of course, much more significant than the number of people currently ill with the disease is the rate of increase in that number. What is true right now is not really so important as what will or could happen. In a time configuration, the characteristics of development are much more revealing than the status quo. But when we are dealing with a developing situation we often fail to understand that we had better focus on how that situation unfolds rather than on its status at the moment.

At the time the article was written, the rate of increase for AIDS was about 130 percent per year. Figure 22 graphs the development of AIDS in West Germany through 1988. Until the end of 1987, the assumption of a constant growth rate of 130 percent provided quite good prognoses. From the end of 1987 on, though, the rate of growth appears to be lower.

An epidemic can have a constant (or almost constant) rate of growth only in its early stages. Later, the rate must decline, and presently I will explain why. In the case of AIDS we couldn't really know how long this "early" stage would last. If it had lasted longer with its growth rate of 130 percent, we would have seen some frightening consequences: from the 262 cases reported in the newspaper article, there would have been 16,863 AIDS cases in five years and 1,085,374 in ten years! (These numbers are computed from the compound-interest formula.)[4]

As we have come to expect, the compound-interest calculation yields numbers that are far from negligible, and they demonstrate the necessity of always taking the process characteristics of a developing situation into account. But even when we do that we can still miss the mark.

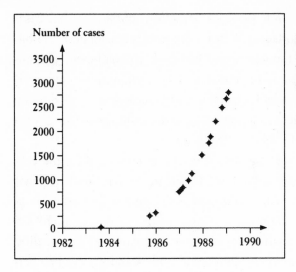

**Fig. 22.** Cumulative number of AIDS cases in West Germany through the end of 1988

In late 1985, two letters to the editor of a German newspaper predicted vastly different trends in the AIDS epidemic: the author of one letter concluded on the basis of his own calculations that "even if the worst predictions come true, fewer people will have fallen ill and died of AIDS between the time we first became aware of the disease and the year 2000 than die of heart disease in a single year." The other letter pointed out that if the worst predictions were to come true the entire human population in its present size, that is, 4.7 billion people, would have AIDS in 2001.[5] Both predictions are as wrong as they are dramatic.

Another newspaper stated at the end of 1985 that only 340 cases of AIDS had been reported in West Germany through the end of November 1985 and that the epidemic was "not [spreading] to the degree previously feared."[6] We can only wonder what prompted that statement. If we recall that, in early September, 262 AIDS cases had been reported, then the increase through the end of November (that is, within three months) was just under 30 percent. That translates into a monthly increase of 9 percent and an annual rate of increase of 183 percent.

Clearly, in these real-world assessments of the spread of AIDS, as in our psychological experiments, people tend to badly misjudge nonlinear growth. Neither the newspaper reporters nor their readers properly understood a development that—at least for a while—proceeded exponentially. Unfortunately, the complications of reality are even greater than simple exponential growth. A closer look at the example of AIDS reveals other things we need to consider when we attempt to interpret time processes in the real world: limited resources, head starts, and transient effects.

## A Premature All Clear?

Thursday, 1 December 1988, was World AIDS Day, and many German newspapers carried optimistic reports: "The Spread of AIDS Has Slowed in the Federal Republic Too," "Education Pays Off: The Number of AIDS Cases Has Dropped Sharply."[7] The first story noted that from the time records on AIDS began to be kept in 1982, a total of 2,668 cases had been reported in West Germany. The doubling time had increased from 8 months in 1984 to 13.5 months in 1988. (It is intuitive, and correct, that longer doubling times correspond to lower rates of growth in the number of cases.)[8]

Reports of this kind are informed by the view that more cautious sexual behavior, awareness of intravenous transmission, the educational efforts of the federal government, people's fears of infection, or all these factors together have produced a deceleration in the spread of AIDS. But that may well be a false conclusion. The AIDS epidemic teaches us to be cautious in our assessments of time configurations and to take many factors into account, including the following major ones. (1) A disease cannot spread exponentially through a limited population indefinitely; its growth rate must drop. (2) Assuming that the first *detected* cases of a disease are actually the first cases can produce an artificially high initial growth rate, which must drop later in the course of the epidemic. (3) Ignoring the length of time between infection and full development of a disease and the variability of this time among individuals can also produce an artificially high initial growth rate, which, too, must drop. Has the AIDS epidemic indeed decelerated?

What we normally understand by "deceleration" is that some quantity becomes smaller over time. What is becoming smaller in the AIDS epidemic? Because the concept "deceleration" is usually not explained in newspaper reports, readers think that the *number* of AIDS cases or HIV infections per time unit is getting smaller, and this is precisely what the second headline I quoted suggests.

But this interpretation of deceleration is by no means justified—not, at least, for AIDS cases.[9] The number of new AIDS cases per time unit has not declined at all. What has declined is the *growth rate*. That means that per time unit the number of new AIDS cases *relative to* the number of existing cases has declined. The growth rate can tell us something about the absolute number of cases only if we know the number of existing cases. The crucial point is that a drop in the growth *rate* by no means indicates a drop in the *number* of cases or infections.

Ten percent is less than 300 percent. But an increase from 10 to 40 AIDS patients (30 new cases) represents a growth rate of 300 percent, whereas an increase from 2,500 to 2,750 patients (250 new cases) represents an increase of only 10 percent. In other words, a low growth rate can go hand in hand with a large number of new cases. This is an obvious point, of course, but many of us need to be reminded of it.

Having clarified what we mean by "deceleration," we can return to our analysis of the course of the AIDS epidemic. We first observe that a deceleration in the growth rate of an epidemic *must* occur and may not reflect any changes either in the infectiousness of the disease or in the behavior of the population. The increasing number of those already infected (who can therefore not *become* infected) accounts for this. Figure 23 shows how HIV infection would spread in a hypothetical population of 1,000 in which 20 percent change their partners each month and the probability that someone in this population living with an HIV-infected person will also become infected is 80 percent.[10]

The curve shown by the solid line in figure 23 reflects the increase in the number of infected persons (measured on the left-hand scale). We see in this simple model that the number of infections increases at an accelerated pace only to slow down as the saturation point is approached. The growth rate (the dotted line "w," measured on the right-hand scale) falls constantly, however. It starts at 16 percent, begins to decline sharply at month 35, and finally drops to 0 at month 80.

The compound-interest formula thus is only of limited use for epidemiological studies because it assumes that every infected individual will spread his infection in proportion to the frequency in which he engages in infectious practices. But in reality that pattern of infection cannot persist over time because as the disease spreads it will happen more and more often that an infected person engaging in an "infectious practice" will be in contact with another already infected individual. This factor alone will cause growth rates in a closed population to decline steadily until they finally reach "0" when the entire population has been infected.

In the seemingly unproblematic calculation of growth rates, however, there is a second pitfall we have to keep in mind if we want to assess declining growth rates properly. Assume that in 1983 someone had discovered a process for diagnosing AIDS and in that year we registered a certain number of cases, say, 16. In the following year, we might have discovered 18 new cases for a total of 34. Recall that the growth rate is the increase over the previous value, expressed as a percentage of that value: here the growth rate is the number of new infections as a percentage of the number of old infections. The growth rate would thus have been $(18/16) \times 100 = 112.5$ percent. Or would it?

The "true" growth rate could be much lower, for the fact that the diagnosing of AIDS began in 1983 by no means implies that no AIDS cases existed before that. They surely did exist but were not recognized, or were only rarely recognized in retrospect, as such. Assume that there were already 100 cases of AIDS in 1982 and the 16 cases discovered in 1983 really represented only the *new* cases. That would mean that the growth rate for 1984 was not over 100 percent but only 15.51 percent

([18/116] × 100). The difference between 15 and 112 percent is not small, and if we mistakenly infer a growth rate of 112 percent when the actual one is just under 16 percent, we have deceived ourselves mightily.

If we take any increase as our initial figure and calculate growth rates on the basis of that number, we will at first grossly overestimate the speed of the process because we will not have taken the process's "head start" into account. An overestimate of this kind probably occurred when researchers first began to study the AIDS epidemic. Once they became aware of the disease, the first cases they discovered were the new ones. They were unable at that point to determine what the head start was. In figure 23 the curve marked "w'''" shows the growth rate if we take the first increase figure as our initial point of reference. We can see how radically this "growth rate" at the beginning of the process differs from the actual growth rate. Not taking the head start into account can explain what seems to be a very sharply declining growth rate.

Thus, a declining growth rate by no means suggests that any precautionary measures or changes in behavior have begun to take effect. In our example, we assigned constant values to the probabilities of changing partners and of becoming infected over the whole time period. Changes in the growth rate reflect changes in behavior only if we can show that these growth-rate changes deviate significantly from the path of decline we would expect the growth rate to normally take and only if we are certain that our calculations are free of a so-called head-start effect.

Finally, to judge the deceleration of the AIDS epidemic properly, we must also take into account that AIDS does not manifest itself immediately after infection. It often takes considerable time before the disease develops fully, and this time varies from individual to individual. Current estimates set an average of eight to ten years. Even if a segment of the population is infected with the HIV virus at about the same time, the infected individuals become ill not at the same time but over a certain period. For example, in a group that was infected in January 1978,

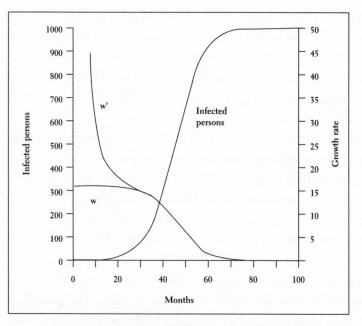

**Fig. 23.** Spread of HIV infection in a hypothetical stable
population of 1,000 individuals

a few people fell ill after a relatively short time. On average, however, it took until 1987 for illness to break out in the group as a whole. And some carriers of the virus did not fall ill until much later. This time range in the onset of illness will initially produce what is known as a "transient effect" on the growth rate of the disease, making it artificially high.[11] This effect comes about because at the beginning of the epidemic's spread the number of cases at a given point in time is made up of those due to fall ill at the average time after infection and those who were infected later but who fall ill earlier than average. For this reason, the number of those ill grows faster at the beginning than it will later.

Recall that the epidemic was known to be decelerating because doubling times were getting longer. In 1984 the number of AIDS cases doubled every 8 months; in 1988 the number doubled every 13.5 months. Exactly how large a deceleration does this change in doubling times imply? If the doubling time increases from 8 months to 13.5 months, then the growth rate declines from 9.05 to 5.27 percent per month. Per year, that is a decline from 182.8 to 85.2 percent and a change of considerable magnitude. What does it mean? Mustn't it reflect changes in behavior? Can it really be simply the rate of decline we would normally expect without any behavior change? I will attempt to settle this question.

The bottom axis in figure 24 represents the time period 1978–92. The small crosses represent the cumulative number of AIDS cases in West Germany as I tracked them at given times in newspaper reports. The value of any cross can be found by multiplying by 10 the corresponding number on the left-hand scale. At the end of November 1989, for example, there were about 2,660 cases of AIDS; in June 1983, there were 43.

The number of cases increased rapidly after 1983. The explanation often proffered for this acceleration was "exponential" growth. But we know that in exponential growth the growth rate remains constant over time. Thus, an epidemic cannot spread exponentially in a limited population, even though in its early stages it may appear to do so. Indeed, if we compare the curve formed by the crosses with curves of exponential growth (shown by dotted lines representing, from left to right, growth rates of 130, 120, 110, 100, and 90 percent per year), we see that it does not match any of them precisely (although a prediction of AIDS cases to date based on a constant growth rate of 110 percent would not have been a bad prediction at all).

These exponential growth rates all show more rapid acceleration than the AIDS epidemic shows. That is true even of the "slow" curves representing growth rates of 100 and 90 percent. Initially these curves rise more slowly than the number of AIDS cases, but then they catch up very quickly and reach almost the same pace as the AIDS epidemic. If we extended these curves farther, we would see that all of them would climb much more rapidly than the curve of AIDS cases. Thus, the AIDS epidemic does not spread exponentially. But knowledgeable people have never claimed it did, because, as previously mentioned, exponen-

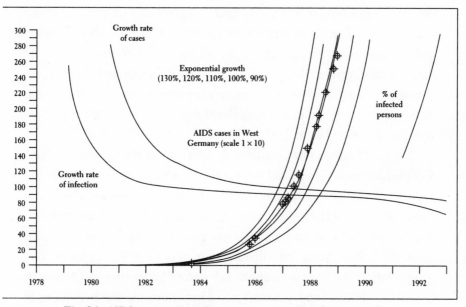

**Fig. 24.** AIDS cases in West Germany as reported in the press and as reflected in the simulation described in the text

tial growth can occur only when completely unlimited "resources" are available. In a limited population the growth of an epidemic must slow.

To judge the deceleration of the AIDS epidemic fully, however, we have to look at other factors as well. We have seen that head-start and transient effects can produce dramatic deceleration in the early stages of an epidemic. How much of the decline in the reported spread of AIDS can be attributed to such effects? Has there been any deceleration, beyond the amount produced by head-start and transient effects, that would reflect the influence of public health efforts?

Making a few initial assumptions that approximate the real situation in 1978, we will chart a purely statistical prediction of the course of the AIDS epidemic in the German population—without removing head-start or transient effects and without assuming any change in behavior or infectiousness—and compare it with the actual data. Assume that in West Germany on January 1, 1978, there were 46 infected persons in an "at-risk" population of 3,000,000, that each month 11.6 percent of the individuals look for and find a new sexual partner, that an uninfected partner who lives with an infected person becomes infected with 53 percent probability, and that the time between infection and illness is on average 96 months. If we use these assumptions to simulate the spread of the AIDS epidemic in our hypothetical population, we get the development shown in figure 24.[12]

Figure 24 shows our simulated growth rate in the number of HIV-infected persons in the population and our simulated growth rate in the number of AIDS cases, both measured on the left-hand scale. (Note that because we based the growth rates on the first increase and did not take the head-start into account, the growth rates are initially quite high.) In June 1983, the annual growth rate of infections is 100 percent. The growth rate in AIDS cases is about 125 percent, a figure consistent with a doubling time of approximately 10.25 months. (At the beginning of 1983 the growth rate is about 140 percent, the doubling time 9.5 months. This latter figure does not quite match up with the 1983 doubling time of 8 months reported in German newspapers. Our figures show the 8-month doubling time occurring at about the end of 1981. But because our numbers agree quite well with the empirical numbers of cases, we will not worry too much about our deviation for 1983, especially since all the figures on AIDS cases in these years are plagued by great imprecision. At the end of 1988, we show a growth rate of about 90 percent for AIDS cases. That figure is consistent with a doubling time of about 13 months, and that, in turn, with widely reported figures.)

As we see, the growth rates are by no means constant but change all the time. They seem to be high at first, which reflects the head-start effect, to level off at about 100 percent, and finally to decline further. It is also interesting and significant that the growth rate for AIDS cases initially remains considerably above that for HIV infections but later runs closely parallel to it. This pattern exemplifies the transient effect.

The solid line in figure 24 that corresponds almost exactly with the curve described by the crosses indicating the actual cases of AIDS gives us our simulated cumulative number of those ill with the disease. This line is the most important of the results. We produced this line using a simulation of the AIDS epidemic that made no allowance for behavior change—the rates of partner change and infectiousness (which is influenced by "safe" behavior) did not vary at any time in the simulation. The deceleration that the solid line illustrates—a deceleration that closely matches that in the actual AIDS epidemic (charted by crosses)—is solely the result of the three statistical phenomena we investigated. Thus the number of people ill with AIDS has to be understood as the result of a process that has undergone *no retarding effects whatsoever.*

Our initial number of 46 individuals infected with AIDS, the size of the population at risk (3,000,000), the partner-change factor of 11.6 percent, and the infectiousness factor of 53 percent are all arbitrarily set parameters. But we should not ascribe unwarranted significance to these initial parameters. Our ability to mimic accurately the actual cumulative increase in AIDS cases in West Germany is important, not because it means that our arbitrarily selected parameters are the ones underlying the development in the real world but because it shows that the number of AIDS cases to date does not necessarily reflect a retardation caused by external circumstances, such as altered behavior in the population at risk. On the contrary, our data support the assumption of a development that has progressed to date without any retardation at all and has been subject only to its own natural rate of deceleration. There is little cause here for comfort.[13]

The one conclusion we should take away from these considerations if we take away nothing else is that we cannot interpret numbers solely on the basis of their size. To understand what they mean, we have to take into account the process that produced them, and that is not always easy.

## Laymen and Experts

So far in this chapter I have described primarily the efforts of individuals who are not professionally engaged in trying to project accurate pictures of future events. But we should look now at predictions made by persons or institutions whose duty or job it is to understand time configurations.

Figure 25 shows several examples of predictions of the number of automobiles in West Germany. The heavy black line represents the actual development; the thin broken lines, the predictions (the year in which each prediction was made is indicated). Through the middle of 1983, the actual development exceeded the predictions significantly (the only exception being the Shell 79/81 prediction). In fact, most of the predictions were proved wrong only a year or two after they were made.

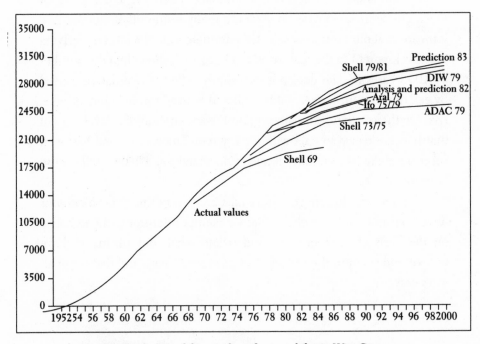

**Fig. 25.** Prediction of the number of automobiles in West Germany
and actual development through 1984

As early as mid-1983 the number of passenger cars in West Germany had reached a figure that the ADAC (*Allgemeiner Deutscher Automobilclub*, the equivalent of the AAA in the United States) only four years earlier, in 1979, was unwilling to predict for the year 2000. Official predictors, like our experiment participants, were apparently not able to foresee this development accurately and vastly underestimated it.

Why? What accounts for these gross errors on the part of people who surely go about solving a problem like this more rationally than our experiment participants, who depend primarily on something like a "feeling"? Why did scientific institutions, just like laymen, underestimate the speed of growth?

One answer lies in the method used to arrive at the predictions. One way experts make predictions is to gather the available data and to find a mathematical function that (with a suitable choice of parameters) fits these data. If they find such a fit, they can accept as their prediction what the selected function yields for future points in time.

They extrapolate, that is, using a model of the process, expressed as a mathematical function. This method bears some resemblance to the layman's method. When our experiment participants set out to predict the production of the tractor factory over time, they too began with a certain model, one that rested primarily on the assumption of linear future development with "corrections for acceleration."

The difference between the lay and the professional predictor consists partly in the range of models available to them. The layman will normally know, and rely most upon, the linear-development model. The professional predictor knows a far wider range of mathematical growth functions and can select the one that seems most appropriate. And unlike the layman, the expert makes this choice *consciously*—not implicitly, on the basis of "feeling" or "intuition."

Why do professionals' calculations sometimes go far wrong despite this advantage? Because even for professionals, feeling still comes into play. If several mathematical functions seem to be almost equally appropriate, which one should they choose? Having chosen a function, which parameters should they use if different parameters yield almost the same results? (At the beginning of the AIDS epidemic, for example, just about any mathematical function that showed a rapid climb seemed to suit the facts: the problem was choosing the "right" function and the "right" parameters.)

These questions reveal the Achilles' heel of the professional's thinking, the weak point where psychology is very likely to come into play. By "psychology" here I mean "feeling" and "intuition." In the automobile predictions charted in figure 25 the feeling that "things just can't go on this way much longer" might well have played a role. Why shouldn't the fact that a mathematician had to spend fifteen minutes hunting for a parking place have some influence on his or her choice of parameters and functions? That the number of cars could continue to grow would seem improbable after such an experience.

What I am saying about professional prediction should not be misunderstood, however, as an attack on prognosticators. I don't know how good or bad business or industry predictions are in general. I want only to call attention to the psychological weaknesses to which even rational professional prediction is prone.

## "Twenty-eight Is a Good Number"

The difficulty we have in gaining a proper picture of temporal configurations necessarily complicates our efforts to deal with them effectively. When we have to cope with systems that do not operate in accordance with very simple temporal patterns, we run into major difficulties.

So far I have given examples only of monotone time sequences in which a development maintains its direction. But it is useful to see how experiment participants deal with developments that show changes of direction, in the form either of oscillations or of sudden reversals. A study conducted by Ute Reichert vividly illustrates the kinds of problems that emerge.[14] The instructions participants received were as follows:

> Imagine that you are the manager of a supermarket. One evening the janitor calls you up and tells you that the refrigeration system in the storeroom for dairy products appears to have broken down. Large quantities of milk and milk products are in danger of spoiling. You rush to the store, where the janitor tells you he has already contacted company headquarters. They have dispatched refrigerator trucks to pick up the perishable goods, but it will take several hours for the trucks to arrive. Until they do, the perishables have to be kept from spoiling.
>
> You find that the defective refrigeration system is equipped with a regulator and a thermometer. The regulator is still operational and can be used to influence the climate-control system and thus the temperature in the storeroom. However, the numbers on the regulator do not correspond to those on the thermometer. In general, a high regulator setting means a high temperature; a low setting, a low temperature. But you do not know what the exact relationship is between the regulator and the cooling system, and you have to find out what it is. The regulator has a range of settings between 0 and 200.

The participants' task was to adjust the regulator to produce a stable temperature of 4 degrees Celsius in the storeroom. But first they had to find out how the regulator settings influenced the temperature.

The key point in this setup is that the climate-control system is not immediately responsive to the temperature in the storeroom but responds to it *with a delay*. If, for example, the room starts out colder than the temperature we have requested (via the regulator), the central system will heat the room, but unfortunately, it will continue to heat the room for some minutes after the temperature reaches the temperature we requested—that is, the system responds to the room's temperature with a five-minute delay. Delays of this kind are common in natural systems, and perhaps the most familiar example for most of us is the residential thermostat that needs a certain amount of time before it brings a room to a constant temperature. At first the temperature is too low, then too high, then too low . . .

Each transmittal of information takes time, and these "dead times" have an important consequence. They create oscillations. If we make no adjustments at all to the regulator in our experiment, the storeroom temperature will behave as shown in figure 26. This figure shows that at a steady regulator setting of 100, which is the initial value given in the experiment, the temperature eventually settles at about 12 degrees.

The experiment participants' job of finding the right regulator setting could be done by experimenting with the regulator and observing the effects of a given setting on the temperature. The most efficient strategy, however, is to leave the regulator at one setting and determine the temperature constant around which the temperatures in the storeroom oscillate; this will be the temperature that corresponds to the regulator setting. We can then change the regulator setting by a certain amount

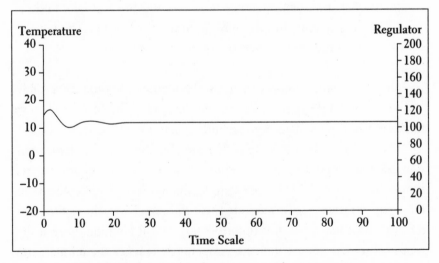

Fig. 26. Temperatures in the storeroom without intervention

and wait until we can determine the new constant. Given the difference between the old and the new constants, we can easily calculate the effect of the regulator.

Because the system oscillates, we have to attend not to individual values but to the temperature constant, or mean value, around which the temperature in the system oscillates. But our participants had great difficulty in coming to this insight.

They tended instead to assume an instantaneous link between temperature and the setting of the regulator. They thought that the temperature they observed after an adjustment of the regulator was the *immediate* consequence of the new setting. This is a false assumption for a system that functions with time delays. But our participants had no end of trouble in shaking off this idea. In the long run, over a period of a hundred minutes, the participants managed to come close to the target value of 4 degrees, but their performance was far from optimal.

The participants assumed that the storeroom temperature would respond to the regulator the way a gas flame responds to a turn of the valve — immediately. With time, however, they learned to bring the temperature in the storeroom to a more acceptable level than it would have reached without their intervention.

Figure 27 ("Examples of participant behavior in the refrigerated storeroom experiment") on page 132 graphs a few different types of behavior. The time scale is marked along the bottom. The temperature curve is measured on the left-hand scale (the dotted line shows the target temperature); the small triangles represent the participant's regulator settings, the numerical values of which are given on the right-hand scale.

Figure 27a shows the behavior of a very good participant. Participant 27a always waits a fairly long time before adjusting the regulator and, as a consequence, slowly develops a feel for the proper setting. He gradually lowers the settings and finally succeeds in bringing the storeroom down to the desired temperature.

Participant 27b is not as successful, although he too finally achieves the proper temperature. He intervenes too often and so deprives himself

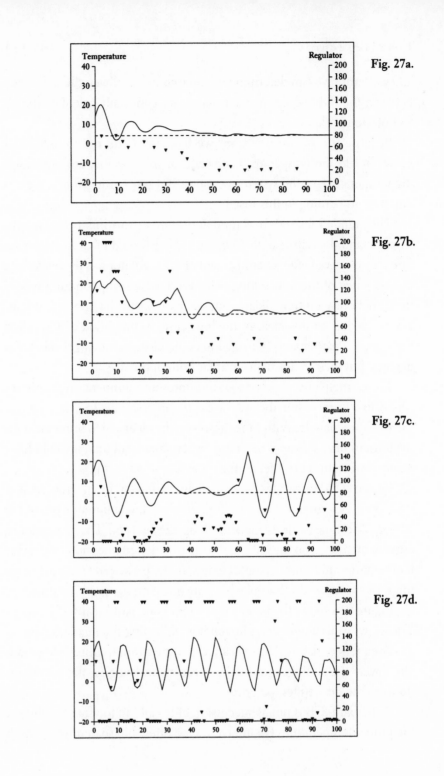

Fig. 27a.

Fig. 27b.

Fig. 27c.

Fig. 27d.

of the opportunity to observe how the temperature oscillations in the storeroom relate to the regulator settings. Finally, however, he hits on the proper strategy and brings the storeroom temperature close to the desired target value.

This is not the case with participant 27c. In the course of the first thirty minutes he displays a "garland behavior" typical of many participants. If the temperature is too high, he turns the regulator down. That lowers the temperature too much and too long, so he turns the regulator up. But then the temperature is too high, and he turns the regulator down again. And so on. We can see on the graph the two resulting "garlands" in the temperature curve between minutes 0 and 30 on the time scale. Next comes a fairly long phase (roughly between minutes 30 and 40) in which this participant does nothing at all. Then comes a phase of frenetic interventions that essentially revert to the "garland" principle. Finally, the participant seems to despair. His settings leap about over the entire range from 0 to 200. Sometimes he selects a very high setting even though the temperature is already high. This suggests that he has, at least during certain periods, ceased to believe the instructions, which told him that high settings would produce high temperatures, and low settings low temperatures. He had apparently started looking for other relationships between temperature and regulator settings, without success, of course. The average temperature deviations in the second half of the experiment are far greater than they would have been if the participant had done nothing at all.

Participant 27d displays an extreme form of behavior. He falls almost immediately into ritualized back-and-forth adjustments between extremes on the regulator scale. He acts on a simple principle: "If the temperature is too low, set the regulator as high as it will go. If the temperature is too high, set the regulator as low as it will go."

This participant's chains of triangles, representing identical interventions, are of particular interest. They mean that the participant keeps resetting the regulator at the same level. That is entirely unnecessary because the regulator cannot change its setting without a participant's active intervention. Also, the participant can see what the current setting is because it is constantly present on the monitor before him.

This participant resets the regulator at 200, for example, even though the regulator is already at this setting and the setting is shown on the computer screen. Why would someone do something that is already done? Probably out of helplessness. The participant has the feeling that he ought to do something because the temperature is anything but optimal. But he doesn't know what to do, so he keeps repeating what he has already done. By doing something instead of nothing, he shows that he has by no means collapsed in helpless surrender to the incomprehensible ways of the system. (Participant 27b displays similar behavior at the beginning of the experiment.)

The difficulties the participants had in coming to a correct understanding of the basically simple laws governing the system emerge clearly in the hypotheses the participants developed about the connection between the temperature and the regulator settings. These hypotheses—derived from tape recordings of the participants thinking out loud during the experiment—break down into three categories. The first and largest category consists of "magical" hypotheses. The participants say, for example, "Twenty-eight is a good number." "A hundred and twenty has something going for it." "Ninety-seven is a good setting." "Odd numbers are good." "You shouldn't use multiples of ten." "One hundred is a good setting; ninety-five is a bad one."

Magical hypotheses are probably the result of overgeneralizations on the basis of local experience. A participant happened to set the regulator at 97, and the temperature, which had just risen, dropped. Given the nature of the system, this intervention probably had little bearing on the effect observed. But the participant is pleased. He notes the "connection" between his setting and the temperature change in the right direction, and he generalizes from this. The numbers are no longer points on a scale for him; they have become individualized and endowed with magical power. In the eyes of this participant at least, they are now "spirits" that breathe a mysterious life into this poor, primitive system.

The hypotheses in the second category look something like this: "Increments of five and ten have different effects." "Changing from increments of ten to increments of one produces an effect." "The key thing is the interval between one and fifty, fifty and one." "Settings of zero, one, two, and three in series are good." "It takes settings of fifty, a hundred and fifty, and two hundred in series to lower the temperature."

These hypotheses are interesting because they reveal a tendency to take the time delays in the system into account. However, it is not really the system's behavior in time but their own behavior the participants are focusing on here. It is not the individual intervention that has meaning but the sequence of interventions. This hypothesis is wrong because a single setting of the regulator—at 23, in fact—would have brought the storeroom temperature to the target value of 4 degrees. But, in any case, these hypotheses do at least address process and therefore touch on an important aspect of the situation.

These sequential hypotheses arise from the "conditionalizing" of simple hypotheses. If a participant has once convinced himself that "ninety-five is a good number" but sees later that the deviation from the target value is increasing with the regulator set at 95, then he is by no means forced to deny the benevolence of the number 95. All he needs to do is introduce some special conditions under which it remains a good number. He can assume, for instance, that 95 will be good only if one proceeds toward it in increments of one.

This process can be extended indefinitely. If it should turn out that adjusting the regulator in steps of one to 95 does not work, then we could hypothesize that we should begin the sequence at 80 and progress from there to 95 by single steps. If this, too, should turn out to be wrong, then we can still rescue our previously erected structure of hypotheses by adding more conditions yet: before we begin the single-step progression from 80 to 95, we first have to set the regulator alternately at 1 and 50 three times in a row.

And so on!

This kind of "progressive conditionalizing" lets us maintain hypotheses indefinitely, and the resulting structure of hypotheses becomes, of course, ever more complex and ungainly. But this, too, has its advantages. If, for example, a certain sequence of steps does not produce the desired effect that strict adherence to the rules should produce, the complexity of the rules allows us to blame the failure on a mistake we must have made at some point in the ritual. In this way we can continue to believe that our ritual is altogether adequate for solving the problem. All we have to do is execute the ritual correctly. At this point, our actions are almost completely divorced from external conditions. We no longer pay any attention to what is happening in the outside world. All that matters is the ritual.

Our participants developed rituals of this kind only in the early stages of the game. The frequent and telling feedback they received on the effects of their actions prevented them from spinning out any elaborate rituals. But in situations where feedback is not frequent and where the intervals between action and feedback are longer, we can expect ritualizations to wax luxuriant. The seeds of such opulent growth are clearly evident in the second category of hypotheses here.

In addition to these two modes of hypothesis formation—generalizing from local experience and the progressive conditionalizing of local experience to the point of ritualization—there is a third mode: "You have to set the regulator high to lower the temperature." "High settings produce low temperatures."

The participants who voiced these hypotheses no longer trusted the instructions or the experiment director but suspected instead that they were the victims of some malevolent deception. They may have been told that low settings of the regulator would produce low temperatures, and high settings high temperatures, but now they had seen through the whole rotten setup. Just the opposite was true!

Hypotheses of this third sort are in a sense "metahypotheses." They imply a revolution in the participant's worldview and call into question the entire structure of the experiment. Trying to establish some presumed link between the regulator and the storeroom temperature is a total waste of time. These participants have found out that there is no such link. Their job now is to get to the bottom of this fraud that the instructions have perpetrated on them. And if the instructions are bogus, then everything else may be bogus, too, like the claim that the regulator still exerts some kind of control over the temperature. This kind of all-encompassing doubt about the very nature of the experiment may well culminate in a thesis reflecting the participant's sense of both liberation and resignation: "The regulator settings have no effect whatsoever on the temperature." Once participants have established that, they needn't trouble themselves any further with the problem.

Bear in mind that the participants in this experiment were working under relatively good conditions. They received running reports on the temperature in the storeroom and could intervene at will. In the real world, systems rarely have such short lag times and rarely provide us with complete information on their behavior over time. This means that the tendencies we observed in the storeroom experiment will be much more pronounced in real situations. In the real world, people tend even more to overgeneralize from local experience, to ritualize, and to believe that no rationally comprehensible principle is at work and that they are the dupes of some mean-spirited practical joke.

## Predators and Prey

Oscillation is one way in which time sequences can change direction. Another equally common phenomenon is a sudden reversal in the direction of a development over time. Growth in an economy is interrupted by a recession. It is suddenly impossible to move a product that has sold well for years. A stream that has always maintained its water level suddenly dries up. A plant that has been growing well for a long time suddenly dies. The stock-market crash of 19 October 1987 is a good example of such a sudden change in direction.

There are, of course, reasons for these shifts and turns. In the fields of ecology, biology, and economics, we find systems that are well buffered. They can absorb a lot of abuse. But at some point, too much is too much, and a liver that has been absorbing too much alcohol for years goes on strike.

"Catastrophes" seem to hit suddenly, but in reality the way has been prepared for them. Unperceived forces gradually eat away at the supports necessary for favorable development until the system is finally unable to resist any longer and collapses. Predator-and-prey systems are simple and readily comprehensible examples of well-buffered systems that undergo seemingly "sudden" shifts in direction.

Biological systems in which a predator (a lynx population, for example) lives on a prey population (say, caribou) often show cyclical development. If conditions are favorable for the prey animal, its population increases rapidly. That in turn creates favorable conditions for the predator population, which can also increase rapidly. But now, of course, things are tough for the prey population, which loses many of its numbers to the large predator population. By depleting the prey population, however, the predators destroy their sustenance. The resulting collapse of the predator population improves conditions for the prey animals, and their numbers rebound; the whole cycle begins again.

Such simple relationships between a single predator species and a single prey species are fairly rare in nature, but they do occur in certain sparse environments. The populations of lynx and caribou in Newfoundland, for example, behave more or less the way I have just described.[15] In richer environments, it is much less common for a predator to depend on only one prey animal; therefore, population patterns are not as clearly cyclical there as they are in the food-poor subarctic regions.

How can participants in an experiment gain insight into such patterns? Can observation provide them with a reasonably accurate picture of such developments? At the University of Bamberg, Walpurga Preussler devised an experiment to help answer this question.[16] The description she gave her participants was as follows:

For centuries the Xerero tribe in Africa has inhabited a fruit-ful region surrounded by desert and bordering on the Tibesti mountain range. Millet is the Xereros' main food. For trade goods, they produce beautiful, artfully woven wool carpets. The Xereros are a very prosperous and peace-loving people. Their re-ligion prohibits them from shedding the blood of either humans or animals. The flocks of sheep that provide them with the raw material they need for their much-sought-after trade goods cur-rently number 3,000 animals. The sheep, which are free to graze year round on the Xereros' territory, are herded together once a year to be shorn and counted. The number of sheep has re-mained roughly constant for many years. But recently the flocks have begun to suffer attacks from hyenas, which now number about 450. Because the hyenas are threatened neither by the Xereros nor by any natural enemies, their numbers can increase unimpeded.

The participants' principal task was to predict over thirty-five time units the curve that the predator (hyena) population would describe (some participants were also asked to predict the prey—sheep—vari-able). The participants had to predict for each time period the next value for the number of predators; after each prediction, they were im-mediately told what the correct value was.

To test the usefulness of certain kinds of additional information, the participants were divided into four groups. One group was asked to pre-dict both the predator and prey variables and to record their predictions on a prepared chart. The second group predicted both variables but did not chart them. The third group predicted only the predator variable and charted it. The fourth group predicted the predator variable but did not chart it. If either paying attention to the prey variable or keeping a chart of the history of the predator variable was helpful in predicting the predator population, this effect would be revealed by differences in per-formance between the groups.

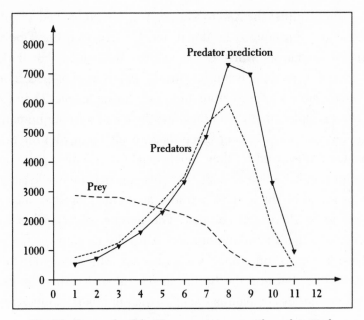

**Fig. 28.** First cycle of the Xerero experiment with predator and
prey variables and with the participants' average prediction values
for the predator variable

Figure 28 charts the most interesting result of the experiment—the
predictions for the first cycle of the two populations. The figure shows
the development of the predator and prey populations and the average
values of all the participants' predictions. On average, the participants
underestimated the initially exponential growth of the predators, but
this is only to be expected, considering the results I outlined earlier in
this chapter. The underestimates are relatively small because the participants were given immediate feedback on the correct number after
each of their predictions.

We can also see from this figure that the participants did not have a
"feel" for the fact that the growth of the predator variable would end at
some point. Instead, they clung to the trend established in their predictions, even at the critical transition from point 7 to point 8.

Indeed, we even find an acceleration in the prediction values. From point 5 to point 6, the participants predict on the average an increase of 40 percent; from point 6 to point 7, an increase of 42 percent; from point 7 to point 8, an increase of 44 percent. The reason for this acceleration in the predictions is clear from the figure. Because each prediction came in below the actual figure for the predator variable, the participants try to make up that difference in their next prediction.

The participants' prognostic behavior can be summarized in the following simple model:

> Take the initial and final values in the process you have already observed and connect those values with a straight line.
>
> Extrapolate this straight line beyond the final value and up to that point in time at which you have to make your next prediction.
>
> Add or subtract the amount of your under- or overestimate from your last prediction. This will give you your next prediction.

As figure 28 shows, the participants were completely surprised by the reversal of the predator variable, just as most stock-market observers were surprised by the crash of 1987. But the participants' reaction was considerably more astonishing because they had a far clearer picture of conditions in the Xereros' territory than any observer could have had of the stock market. Once the sudden deceleration in population growth seen in period 8 signaled impending catastrophe, the participants did anticipate further decline but were surprised once more at its rapidity (period 9).

Even more striking are the results shown in figure 29. The predictions were split into two sets and the averages for each set were plotted. One curve reflects the values of all those participants who were asked to predict only the predator population. The other reflects the values of all those who were asked to predict both the predator and the prey populations.

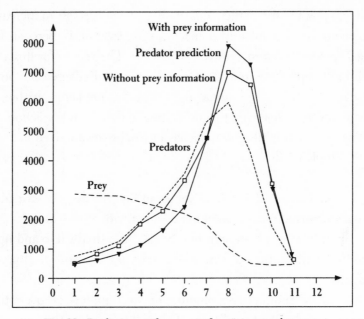

**Fig. 29.** Prediction performance of participants who were
given the "prey" prediction and those who were not

We would expect the participants who witnessed the steady decline
of the prey population to draw some conclusions that would be reflected
in their predictions of the predator population. There are fewer and
fewer sheep and more and more hyenas. At some point the hyenas will
start going hungry too. Surely the participants ought to have noticed
that the prey population shrank significantly during periods 4 and 5.
And if the prey population declines, that must eventually affect hunting
opportunities for the predators.

But if any of the participants came to such conclusions, they
made no use of them whatsoever in their predictions. On the contrary:
the "predator-only" participants seem to have done better than the
"predator-prey" ones. They did not overshoot the catastrophe point by as
much as the predator-prey group did. (The difference is statistically significant.)

How can we explain this? Was too much asked of the predator-prey group? They were constantly receiving more information than the participants in the predator-only group. Did the additional chore of integrating the two developments produce an informational overload? And did that in turn induce them to adhere rigidly to their automatic mode of prediction? It would appear that the predator-only participants, less burdened with information, had enough free attention left over to anticipate the deceleration in the predator growth rate.

Requiring half of the participants to record the growth curves of the variables on a graph—to convert information received over time into spatial information—was slightly beneficial. These participants came closer to predicting the first "catastrophe"—the reversal of the predator variable—than did the participants who did not graph their curves on paper. In other words, converting "time" into "space" seems to help people comprehend temporal configurations, though in this case it did not help much.

Neither aid—neither the reports on the actual development of a closely related variable nor the recording of curves in graphic form—had much effect on the results. The experience the participants gained in the first cycle of the experiment, however, did influence their prognostic behavior. The continuation of the experiment through a second cycle showed improvement. Figure 30 charts the average course of the predictions for the entire experiment, over thirty-five time units. In the second cycle, the participants predicted the growth of the predator population better than they did in the first cycle, and the second reversal of the predator population did not catch them as much off balance as the first did. Over the course of the experiment, they clearly became more attuned to the processes involved.

The improvement these subjects showed might induce in us a sense of optimism: perhaps human beings can learn how to deal with time configurations after all. But we have to remember that participants in this experiment were working under nearly optimum conditions. They had only one task to perform and were not distracted from it by other tasks. Some of the participants received information about

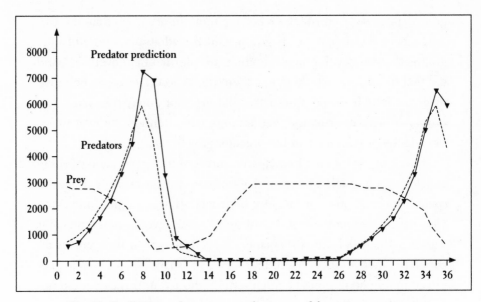

**Fig. 30.** Prediction performance over the course of the entire experiment

the prey population. After each prediction, all the participants received feedback on the accuracy or inaccuracy of their predictions and were told the correct value. In the real world, these are all totally unrealistic conditions. The information a newspaper-reading citizen receives about economic developments or the spread of epidemics, for example, lacks both continuity and constant correctives. Information comes in isolated fragments. We can assume that those conditions make it considerably more difficult to develop an adequate picture of developments over time.

## The Moths of Kuera

Kuera is a village in the Nile delta. Both cotton and figs are grown there. The cotton crop depends on the activity of the death's-head moth (*Acherontia atropos*), which pollinates the blossoms of the cotton plant. But this same moth damages the fig crop by boring holes in the figs to feed on the juice, not only destroying fruit mass but also opening a passageway for pests that produce rot in the fruit.

So on the one hand the death's-head moth is essential, but on the other it is a pest. What is clearly desirable in this situation is to limit the moth population to the absolute minimum necessary for the cotton crop.

A certain species of predatory wasp that feeds on the moths can be used to control their population. This wasp species is quite simple to manage. It builds its "papier-mâché" nests, which are readily visible, in trees, and at night, when the wasps are asleep, the nests can easily be collected. Also, the nests can be stored and then set out again at will. (Zoologists and botanists will have to forgive us for our less-than-strict adherence to zoological and botanical reality; accuracy was not our top priority here. We wanted to create a system that would mimic a predator-prey system, and we also wanted to present it in wording that would allow participants to think of the experiment as an exercise in biological pest control.)

The participants' task was to collect wasp nests and set them out again in such a way that the moth population would remain as close to a designated target level as possible. This target level is represented on the following graphs by the 400 level. (An actual moth population would always be much larger than the values shown on the charts, but for the sake of surveyability we have kept the scale small.)

The collecting and dispersing of wasp nests takes time. Therefore, no given measure takes effect until two months after it is ordered. The participants are made aware of this time lag, and they are asked to conduct their control effort over thirty-seven time periods ("months").

Figure 31 graphs the activity of a participant who managed this task superbly. The upper part of the diagram shows the changes in predator and prey numbers, that is, the changes in the populations of wasps and moths. Both populations are nearly stable after fourteen months, or a total of only four interventions. The interventions are charted on the coordinate system below the main graph. Each triangle represents an

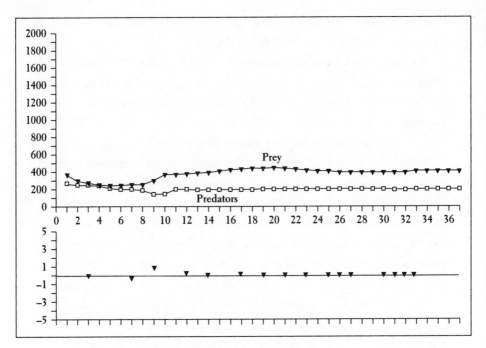

**Fig. 31.** Results of participant 13 mbw in the Kuera experiment.

intervention. A triangle above the zero line indicates a dispersal of wasp nests, a triangle below the line a collecting of nests. The distance of a triangle from the zero line indicates the intensity of the intervention.

Participant 13 mbw proceeds very calmly, waits for a long time, observes. The steps he takes are relatively small, but his minimalist approach is obviously appropriate. From the very outset this participant focuses on understanding developments. That is no simple task, because the moths tend to increase to the maximum level determined by the available food. And once the wasps are set out in the environment, they, too, increase independently. Also, quite apart from the activity of the participant, a small population of wasps migrates into the Kuera region and slowly increases. The system is characterized, in other words, by internal dynamics of some significance.

If we look at month 9, we see that this participant has understood these internal dynamics quite early on. Our participant distributes wasp nests here even though the moth population is still below the target value of 400. He has understood that the moth population increases rapidly, and so he intervenes early enough to check that development. His main goal is to keep the predators at an optimum level because he has hypothesized (correctly) that, by maintaining a certain constant population of wasps that he has not as yet determined, he can hold the moth population at the target level.

By about month 15 this participant has roughly determined the range in which he needs to keep the wasp population, and his later interventions amount to fine-tuning the system. With minor adjustments in the number of wasp nests, he can control the moth population very precisely, and the number of moths deviates less and less from the target value. (The small scale used in figure 31 makes it impossible to see these minor deviations.)

While running the experiment, we observed that this participant behaved calmly, was mindful of data in his hypotheses, generalized very little, noted the changing numbers of moths and wasps, and attempted to translate into spatial information the information he gathered about the time configurations in question.

Now let's look at the behavior of participant 02 mjg as recorded in figure 32. We can see immediately that this participant had much more trouble with the task than 13 mbw did. The moth population is either zero or far above the target value, and the participant's interventions are far cruder than 13 mbw's. The outstanding characteristic of this participant's behavior is that he reacts to the *current* status of the predator and prey populations and ignores the time-related aspects of the problem (population growth and decline, acceleration, and retardation). Never during the entire course of the experiment does he understand that he is dealing with a process in time. He remains a captive of the moment.

He notes at first that the moth population is declining, so he cuts back on the wasp population. The result is an almost explosive growth in the moth population. Then in months 6 and 7 the participant attempts, too late and too timidly, to stem this tide. When these measures

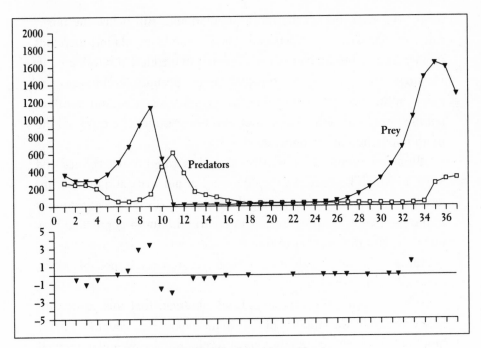

**Fig. 32.** Results of participant 02 mjg

fail, he resorts to much too drastic ones in months 8 and 9 and eradicates the moth population almost entirely. In months 10 and 11, having just dispersed a large number of wasp nests, he takes them in again. (This measure has no effect on the moth population, at this point, and the wasps, deprived of their main source of nourishment, would starve relatively quickly anyway. On the other hand, this measure is kind to animals and ensures an adequate supply of wasps.)

As early as month 8 this participant develops a conspiracy theory: "No fair! This computer is cheating on me!" Then, about midway through the experiment, he starts observing trends. The gradually rising population of moths even moves him to take preventive action in month 18. Acting too timidly once again, he disperses a very few wasp nests.

He never understands that he is dealing with *processes*. In month 22 he vacillates between dispersing wasp nests and gathering them in because, on the one hand, he fears the rapid population growth of moths but, on the other, the moth population is still way below the target value.

Next, he drastically underestimates the obvious and almost exponential growth of the moth population from about month 22 onward, and in months 25–29 he tries, much too late, to counter it. And once again he expresses his conspiracy theory. "Forces of evil" are at work behind the scenes: "The computer is cheating!"

The moth population climbs far above the target value, and our participant reacts much too drastically. If the experiment were extended, we would see the moth population drop to zero again.

At the end of the experiment, participant 02 mjg remarked: "No matter what you do, you can't do anything right."

In figure 33, participant 17 mtm presents a similar picture. Here, in months 3, 4, and 5, the participant disperses large numbers of wasp nests even though the moth population is clearly declining. No sooner have

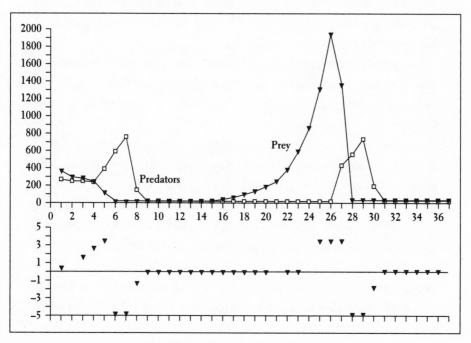

**Fig. 33.** Results of participant 17 mtm

the nests been dispersed than our participant frantically gathers them in again in months 6, 7, and 8, and the participant is left from month 9 on with neither moths nor wasps.

The moth population gradually rebounds, and our participant, now in a mood of aggressive helplessness, gives himself up to blind activity and in months 9–20 collects every wasp nest he can lay his hands on. He even takes in a small number of wasps that migrate into the Kuera region. This measure is totally unnecessary to promote the moth population, which has already begun to rebound, and even harmful because the "immigrant" wasps would have slightly retarded the growth of the moth population.

As a result, the moth population explodes from about month 18 onward, but our participant, failing to notice this until month 23, persists in his aggressive antiwasp policy. In months 25–27, he disperses wasp nests in massive numbers, a step that quickly brings about a total collapse in the runaway moth population. Having dispersed the nests, however, our participant frantically gathers them in again in months 28 and 29, and from month 31 onward we find the same situation we saw back in month 9.

This participant learns next to nothing during the entire experiment, gives himself up increasingly to a mood of aggressive helplessness, and indulges in ineffectual routine measures so that he can feel he is doing something.

Figure 34 charts the activity of a participant who, like participants 17 mtm and 02 mjg, begins with a strategy of aggressive countermeasures. If there are too few moths, he collects wasp nests. If there are too many moths, he disperses wasp nests. He maintains this strategy with its relatively extreme interventions and abrupt reversals until about month 12. Then this participant shifts to genuine process control and tries to act with foresight. The measure he takes in month 13 exemplifies this change. He disperses only a small number of wasp nests even though the moth population is still much larger than target value. Our participant has hit on the idea of determining what he calls a "zero value."

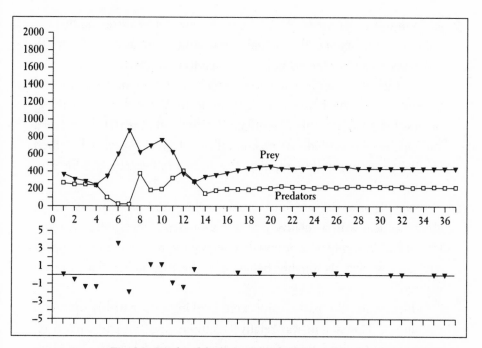

**Fig. 34.** Results of the "educable" participant, 04 mlg

What he means by this is a value for the wasp population that will keep the moth population constant. As the further course of the experiment shows, our participant is highly successful in determining that value. With relatively few and well-tempered measures, he adjusts the moth population to the target value. Participant 04 mlg is an example of a person who learns from his mistakes and can alter his behavior. And this change can take place quite abruptly. Here the change in strategy occurs in month 12 or 13.

The participants in this experiment display a wide range of behaviors. Participant 13 mbw is clearly the star—but also the exception. The other participants, on average, learn very little from the experiment, and have great difficulty developing a control strategy. The behaviors of 17 mtm and 02 mjg are extreme cases but still display characteristics of average behavior.

All the counterproductive behaviors we have seen before turn up again in this experiment. We find massive countermeasures, ad hoc hypotheses that ignore the actual data, underestimations of growth processes, panic reactions, and ineffectual frenetic activity.

On the other hand, as in the storeroom experiment, we had a participant who sat back and waited, trying to understand the underlying process before attempting to regulate it. Even better, perhaps, we had a participant who started out making all the typical mistakes but then learned to observe and manipulate the process rather than just the situation at each moment.

Overall, however, we must conclude that most participants' native ability to deal with a relatively simple time-dependent system is minimal. It would surely be worthwhile to equip the participants with better strategies so that reasonable behavior is the rule rather than the exception.

Once again, the participants were working under nearly optimum conditions. The system was relatively simple, and the participants received complete and correct information without delay. The lag times in the system's reactions were relatively short and surveyable. Under other conditions our participants would probably have found their task even more difficult. If the system had been more complex and the lag times longer, if information had been slower in arriving and been incomplete and partially wrong as well, and if the interaction between the variables had been more complicated, our participants would no doubt have run into greater problems.

The system in Kuera made no superhuman demands of the participants. The problems it poses are quite easy to solve. How? The solution is nothing arcane. All it requires is keeping a few utterly simple rules in mind: Try to understand the internal dynamics of the process. Make notes on those dynamics so that you can take past events into account and not be at the mercy of the present moment. Try to anticipate what will happen. Elementary, my dear Watson!

# Six

# Planning

I f we want to deal rationally with a complex problem, the first thing
we do (tentatively, at least) is define our goals clearly. Then we construct a model of the specific reality or modify an existing model. We
may have to observe the system for quite a while to understand the connections between its variables and need to gather information on the
present state of the system so that we know how it is behaving now and
how it is likely to behave in the future. Once we have done all that we
can move on to the planning stage.

## "Go Make Yourself
## a Plan . . ."

What is planning? In planning we don't *do* anything; we just consider
what we *might* do. The essence of planning is to think through the consequences of certain actions and see whether those actions will bring us

closer to our desired goal. If individual actions will not achieve our purpose, we have to lay out sequences of actions. "First I'll move my pawn to D4, where together with the bishop it will protect my queen, and then with this knight . . ." This is what a brief segment of a chess player's planning process might look like.

Planning consists of examining the consequences of individual actions, then of stringing individual actions together into sequences and examining the possible consequences of these sequences of action. We do this in our heads or on paper or with a computer. Planning is a matter of sending up mental trial balloons. We ask ourselves, "What will happen if I carry out step A? And what will happen if I then add step B to step A?"

In planning, we develop more or less long chains of imagined actions. These chains consist of individual links that, if they are complete, each comprise three elements: a *condition* element, an *action* element, and a *result* element. "Given such and such conditions, I could take this or that action and achieve this or that result." This would be the complete form of a single unit within a planning sequence.

Plans can branch out in different directions. It may occur to a planner that a certain action may lead to not just one result but different results, depending on a priori conditions unknown to him. "And if my opponent doesn't protect her queen with her bishop but instead moves her rook onto A5? . . . Then I can take her knight with my bishop and then . . ." This is an example of a plan that can follow different paths depending on circumstances.

Planning sequences can also include loops. If, for example, we suspect that a certain action may sometimes produce the desired result and sometimes no result at all, we may say to ourselves, "And if it doesn't work this time, I'll just try it again."

A planning process can take on a more or less complicated form such as the one shown in figure 35. This is a branching structure that originates at point $S_\alpha$ and moves toward point $S_\omega$.

$S_\alpha$ is the starting point of the planning process and $S_\omega$ its final goal.

(There are planning processes that have several starting points and work toward several goals, but I don't want to make things complicated here.) The arrows represent actions, and the black dots the arrows point to represent the goals we hope to achieve with those actions. A "fork" represents an action that may have more than one result; a "loop" represents an action that may need to be repeated.

Figure 35 shows just about all the eventualities that can occur in a planning process. It also shows that we can distinguish between two kinds of planning: forward planning and "reverse" planning. In forward planning, we begin at the beginning; it is, in some sense, the "natural" form for planning to take. We plan the way we will actually act—forward. Reverse planning is thus in this sense "unnatural" because we can't act in reverse.

But even if we can't act in reverse, we can plan in reverse. We can quite easily consider what conditions would have to prevail just prior to the desired goal in order for us to achieve that goal by means of a specific action. For example, if we want to travel by train from Chicago to Boston, the best strategy isn't necessarily to investigate train connections from Chicago to Pittsburgh, from Pittsburgh to New York, and from New York to Boston and then develop our travel plans on the basis of that information. We can instead find out which trains come into

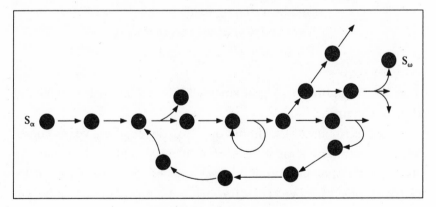

**Fig. 35.** Planning structures

Boston from the west during the period when we would like to arrive. We can then find out when these trains leave from, say, New York or Buffalo, and in this way we can plan our way "back" to Chicago. And we can, of course, combine both kinds of planning, shifting back and forth between the forward and reverse modes.

For reverse planning it is crucial to have a clear goal in mind. If the goal is foggy and unclear, we have no solid frame of reference for the question "What single action will result in the desired goal?" This is yet another reason we should be rigorous in clarifying our goals.

The fact that goals are often unclear may well explain why people show little spontaneous enthusiasm for reverse planning. But even when people do have clear goals and could use reverse planning effectively, they rarely do. When I studied a group of people asked to do proofs in formal logic, I found in 1,304 recorded units of the participants' thinking out loud not a single example of reverse planning.[1] For mathematicians and logicians, however, reverse planning is often a standard method.[2]

Theoretically, planning is a simple matter. In our minds, we lay out a sequence of actions, combine forward and reverse planning, and eventually find our way from the starting point to our goal. Or do we?

> Go make yourself a plan
> And be a shining light.
> Then make yourself a second plan,
> For neither will come right.

sings Peachum, Brecht's beggar king, in his "Song on the Inadequacy of Human Enterprise."

As a description of Peachum's own experience, these verses constitute a gross understatement. Most of his plans go off like clockwork. Indeed, shortly after proclaiming his philosophy, Peachum lays out for London's chief of police, Tiger Brown, a vast array of plans that he will execute if Brown tries to bar London's beggars from the coronation ceremony. Brown is so impressed by the thoroughness of Peachum's plans that he soon lets himself be enlisted in Peachum's plan to capture Macheath. This plan, too, works beautifully, and had the king's messenger not arrived with a last-minute reprieve, Macheath would have perished on the gallows. But, alas, we are not all Peachums. What is it that makes planning so difficult for us?

One problem is that only rarely and in relatively uninteresting areas can we plan completely. Ultimately, planning obliges us to investigate a sector of reality—a "problem sector"—to determine the possible transformations we may be able to effect there.[3] But because even moderately complex realities, such as puzzles, offer a multitude of transformation possibilities and because both forward and reverse planning processes can branch out so widely, a complete investigation of an entire problem sector is utterly impossible.

As we all know, chess is a sector of reality with a limited number of configurations. Therefore, the number of moves and countermoves is also limited. In theory, then, we should be able to plan a move in chess—indeed, even an entire game strategy—"completely." But in fact we can't, because the number of possible transformations is so large that no one, not even a computer, has yet been able to sort through and evaluate them all.

Because the vastness of problem sectors prohibits us from investigating them completely, we must narrow our focus. The psychology of problem solving has identified many heuristic devices for accomplishing this. One common focusing procedure is "hill climbing"—considering only those actions that promise a step toward our goal or, in other words, that diminish the difference between what is given and what we want to achieve. This would seem to be one we could take for granted. But it is tricky to use because as we come closer to our goal—the summit—we may find ourselves on a secondary peak rather than on the main one. At some point during the ascent the way to the secondary peak may be steeper than the one to the main peak, and if we apply no method other than hill climbing, we may wind up on the wrong path.

It can be dangerous to rely entirely on the steepness of the climb—on our increasing closeness to our goal—as a criterion of success. Nevertheless, hill climbing does keep us from taking any detours, because they always represent a temporary deflection from the direct path to our goal.

Another way to limit problem sectors is to look for "intermediate goals." We can develop intermediate goals by using reverse planning. We can also choose as intermediate goals situations that offer a variety of possibilities for future action even though we cannot see any direct path leading from them to our goal. An example of this kind of situation in chess is a good placement of pawns or control over the four central squares on the board. These "favorable" situations are ones that offer high "efficiency diversity" situations from which one can move efficiently in many different directions.

If we are at a total loss as to what to do, we can act on the basis of what has proved successful in the past. Selecting actions on the basis of their frequency of success in the past also narrows our focus. We must not succumb to "methodism," however, choosing an action only because it has often worked before.

My purpose here is not to provide a systematic discussion of all the different ways we can narrow our problem sector. I only want to point out that there are many.

If there are many ways of narrowing our problem sectors, however, which ones should we choose, and when? When should we use a combination of forward and reverse planning? When is hill climbing appropriate? When is the efficiency-diversity method best? When should we study past success? There are answers to these questions. Reverse planning works poorly or not at all if our goals are not clear; the efficiency-diversity method is appropriate then. And only if we are completely uncertain about the structure of the problem sector, and thus about the best path to our goal, should we resort to hill climbing, perhaps interspersed with trial and error to counteract the inflexibility of this approach. We simply need to know when to do what.

Methods for narrowing problem sectors make methods for expanding them necessary too. Narrowing a sector lets us operate in a surveyable field, but the possibility exists that we are in the wrong one. If a reasonable period of exploration leads us nowhere, we need, therefore, to consider changing our field, and there are several ways of doing that.

The simplest of these is *free experimentation*. Trial and error is a means of introducing mutation into a planning process, so that we do not limit ourselves to operations that seem to lead to our goal (hill climbing), that take us to a favorable strategic position, or that were successful in the past. Instead, we try everything conceivable in the given situation. This method is fairly primitive, and it poses problems because we often do not realize we are captives of our old ideas and are by no means considering all the available possibilities.[4]

Another way of expanding a problem sector is *culling unsuccessful strategies*.[5] In this method, we identify the features that previously unsuccessful approaches to a problem have in common, and then we develop new approaches that do not have these features. If what we have tried so far has not worked, we need to cull the features of our unsuccessful actions and replace them with new ones.

This procedure is particularly effective in helping us overcome deeply entrenched patterns of thought. Long association of a particular item with a particular function makes it difficult for us to imagine that item's serving any other purpose.

If, for example, we give experiment participants a candle stub, some matches, and a few thumbtacks in a matchbox and ask them to fasten the candle stub to the wall so that it can be used for "optical experiments," the participants will hit on the idea of pinning the matchbox to the wooden wall with the thumbtacks only much later and with much greater difficulty than if we gave them the candle stub, the thumbtacks, and the matchbox separately. In the first case, the matchbox is subject to a particular association. The participants see it only as a container for the other items, not as a possible wall fixture for the candle.

Perhaps the most important method for expanding a problem sector is *thinking by analogy*. When our participant in the Greenvale experiment compared the manufacture of watches to rolling cigarettes, the clear picture of the production process she formed in this way opened up new planning possibilities for her. In the refrigerated-storeroom experiment, the essential point for participants to grasp was the time lag between an action and the effect produced by that action. A participant who saw an analogy between setting the regulator at a new value and sending bills to customers ("I don't get my money instantly either") may not have developed an earthshaking idea, but he did hit on the one idea he needed to solve his problem.

As with methods of narrowing problem sectors, certain methods of expanding sectors are appropriate in certain situations. Culling unsuccessful strategies is useful as an analytical method of expanding a problem sector after a long period of unsuccessful effort. Free experimentation is appropriate when we find ourselves unable to set any more intermediate goals and realize we are thinking in circles, constantly revisiting the same repertoire of ideas. If we have the sense that we have exhausted the possibilities of a problem sector, we should work toward expansion by analogy.

We can think of planning as a process of narrowing our problem sector, searching through that sector intensively for possible ways of solving our problem, expanding the sector if that search proves unsuccessful, limiting the new problem sector, searching through it, and so on. The effectiveness of the whole process depends on the methods we have available for limiting and expanding problem sectors and whether we know which ones to apply when.

Limit here, expand there—we must keep in mind that this is the approach we should take in theory. Our ability to follow it in practice will always be constrained by both real-world conditions and the demands of particular problems. Time limits may force us to develop only crude plans or may curtail planning altogether. In addition, there are instances in which we should not overplan or even plan at all, regardless of the amount of time. Some situations depend on such a multitude of other processes that the particular details simply cannot be anticipated. As

Hans Grote, a building contractor, observes, a soccer coach will not tell one of his forwards that he can be certain of scoring if, in the sixth minute of play, he approaches the opponent's goal from the right at an angle of 22 degrees and, 17 meters in front of the goal, kicks the ball at an angle of ascent of 10 degrees, 11 minutes."[6]

Detailed planning in this context is a waste of time. We would do better to follow the advice of Napoleon, whose motto for such situations was, "On s'engage et puis on voit!" (Freely translated: "One jumps into the fray, then figures out what to do next.")

In very complex and quickly changing situations the most reasonable strategy is to plan only in rough outline and to delegate as many decisions as possible to subordinates. These subordinates will need considerable independence and a thorough understanding of the overall plan. Such strategies require a "redundancy of potential command," that is, many individuals who are all capable of carrying out leadership tasks within the context of the general directives.[7]

This kind of delegation should not be confused with the unproductive situation described by Graf von Moltke, a renowned military strategist: "If a general is surrounded by a number of independent advisers, the more of them there are, the more eminent they are, and the more intelligent they are, the worse his situation will be. He listens to this adviser, then to that one. He follows his first adviser's essentially sound recommendation up to a certain point, then the even sounder recommendation of his second adviser. Then he acknowledges the validity of objections raised by a third adviser and the cogency of suggestions offered by a fourth. And so we can safely bet a hundred to one that thus equipped with nothing but the best of intentions, he will lose his campaign."[8] Here, rather than independent subordinates working within the context of an overall plan, we have eminent advisers competing to set the general strategy. A "redundancy of potential command" may also be psychologically unsound in some cases, however, because it adds to the complexity of the basic problem the unpredictability of those independent agents to whom we have delegated authority.

In practice, though, excessive delegation is not often a problem. In

another version of the fire experiment, researchers found that participants acting as fire chiefs showed a strong tendency to take away from
the individual units what little independent decision-making capacity
they had.[9] The chiefs preferred to subordinate the units to their own
central command, even though it would have been better in some situations (for example, those requiring prompt action) for the units to act
independently. The chiefs wanted to feel that everything was under
their control, even though that was obviously counterproductive. Other
researchers found that problem solvers working under stress used the
personal pronoun *I* more often than those working in relaxed circumstances.[10] Does this finding suggest a tendency to resort to centralized
regimes in stressful situations?

By identifying situations in which we should plan little, if at all, we
have touched upon one extreme of a broader issue. In planning, as in
information gathering, we are faced with the problem of scale. We can
make plans that are too crude and plans that are too detailed. The trick
is to plan with an appropriate degree of detail. But what is appropriate?

The same kind of planning is not necessarily appropriate for all
spheres of reality. Grote described how he used one method to good effect in planning and then constructing a sewage-treatment plant in
much less time than is usually required for comparable projects. He was
determined to apply the same planning technique to the construction of
a secondary school; the outcome was different, however:

> We found very quickly that our chosen method failed when
> faced with the greater complexity of this building process.
> First of all, its all-inclusiveness, which detailed hundreds of
> operations, placed unreasonable demands on everyone working
> at the construction site; second, the actual construction pro-
> ceeded very differently from what had been envisioned in the
> models, despite the expertise and intelligence that had gone into
> them; third, we had programmed in the worst blunders ourselves
> by miscalculating the number of workers needed in each area.

We didn't know in advance that the contractors and subcontractors—and the engineers in the construction firms—had been thrown way off by the complexity of the project and had themselves made disastrously incorrect estimates of the number of workers needed.

Grote summed up: "Complete planning should be tempered by the concept of 'nonplanning.' "[11]

The difficulty of finding the correct scale for planning accounts for many failures. The more uncertain we are, the greater our tendency to overplan, for example. In a situation we find threatening to begin with, we try to foresee all possibilities and make allowance for every conceivable mishap. This approach can have ruinous consequences. The more extensive our understanding becomes, the more the planning process will impress on us the myriad possible results. Planning, like the gathering of information (for planning, too, is a form of information collection), can increase our insecurity rather than reduce it.

To return to the example of the soccer coach: If the coach is going to determine the positions from which each of his players should shoot, he should keep in mind that damp earth can stick to soccer shoes. And a clump of dirt between shoe and ball can play havoc with the angle of the planned shot. It would therefore be wise to study the average size of clumps of dirt and their frequency of occurrence, as well as the places on a soccer shoe where they are most likely to cling. But then if we consider that soccer fields in the north tend to be sandy while those in the south have a more claylike consistency, we have to . . .

No one would ever go to such ridiculous lengths, you say? Oh, yes, they would!

A participant in the Greenvale experiment was drawn to the problems of the elderly, and he determined that something had to be done to foster communication between elderly citizens and their children and grandchildren. His interest prompted him to ask about the presence of telephones in Greenvale's nursing homes. But because not all of

Greenvale's elderly lived in nursing homes, our participant had to investigate the availability of private and public telephones. He obtained from the experiment director the location of all the telephone booths in Greenvale, entered them on a map of the town, and then set about calculating, with the help of a ruler and a pocket calculator, the average distance an average senior citizen would have to travel to an average telephone. On the basis of this information, he would plan the location of new telephone booths.

He had, of course, long since lost sight of Greenvale's elderly citizens and their needs. They served only as a pretext that let him seek refuge in a Lilliputian planning project and escape his helplessness and the complexity of real decision making. Looking closely at a problem often increases our insecurity, and a retreat into a minuscule but detailed planning process can help us feel we are applying the full force of our rational powers to the uncertainty of the situation while letting us put off the evil day of action—after all, we have to plan carefully before we act. Thus, our diversionary planning process lets us off the hook for the time being and spares us the possibility of failure.

The sequence "insecurity" → "precise planning" → "greater insecurity" → "even more precise planning (preferably in a familiar field)" → "dim awareness that one is not coping with the real problem" → "refusal to make a decision" need not end at the last step. A helpless, hapless planner may drag the process out further, burrowing deeper still into an ever more refined and more narrowly focused planning process before breaking out in a "liberating act" that may be completely at odds with all the planning and amount to no more than blind, random activity.

Even if overly detailed planning does not produce greater insecurity, it can have dire consequences, for if we believe we have anticipated, and prepared for, all eventualities and things *still* go wrong, we suffer a much heavier blow, one that undermines our confidence much more than if we had approached our problem with Napoleon's motto in mind, expecting that something would surely go wrong but that we would find a way to deal with problems as they emerged. If we expect the unexpected, we are better equipped to cope with it than if we lay extensive plans and believe that we have eliminated the unexpected.

The attempt to escape the uncertainty of a complex situation can take the form of flight to the high ground of minute, detailed planning, preferably fortified by mathematical formulas, for what we arrive at by calculation is bound to be right. Perhaps this explains the great popularity of formal methodologies and "mathematization" in the "imprecise" sciences of psychology, economics, and sociology?[12] Obviously mathematics is not responsible for its misuses, and an idea expressed clearly in mathematical terms is preferable to one expressed unclearly in lay terms. What we do have to avoid, however, is reducing and simplifying reality until we are able to fit it into a predetermined formal pattern; ideas altered in this way no longer reflect the original reality from which they are derived.

Crude planning is, of course, just as dangerous as overly detailed planning. I don't stress it, though, because it comes in for criticism more often than overplanning does. Nevertheless, the psychological determinant that prompts us to plan inadequately or not at all is probably the same as the one that seduces us into excessive planning. In both cases, the planner faced with the need to make a decision is driven by insecurity.

## Rumpelstiltskin

I have defined a planning unit as consisting of a condition element, an action element, and a result element. (The result element is, of course, an "expected result" element; what actually results from an action may not be what the planner expected.)

If we plan conscientiously with these three elements in mind, our task will be demanding. Certain actions require certain conditions. If we want to carry out an action, we must first create the necessary conditions if they don't already exist. The execution of actions takes time and effort; we must take that into account too. And if the result is not exactly what we wanted, we may have to undertake additional action. In sum, planning is much easier if we ignore the condition element and assume that our action is *generally* applicable, if we ignore difficulties inherent in the action itself, and if we assume that the action will produce the desired results.

Because planning only involves imagining our actions, we are essentially free from the irksome conditions of reality, and nothing prevents us from simply ignoring the conditions necessary to carry out an operation. Since we human beings tend to think in the abstract anyway, ignoring those conditions comes quite easily.

Rumpelstiltskin, you may recall, is something of a planner. Today he'll bake, he says, tomorrow he'll brew, and the day after that he'll take the queen's child. His plans go awry when he neglects to keep his name to himself. This kind of planning is quite common. "Today I'll enter my data into the computer, tomorrow I'll evaluate them, and the day after tomorrow I'll write the concluding chapter of my thesis." But what if all the university computer terminals are occupied? And what if one becomes free but the center closes early? Perhaps the instructions for running the evaluation program have been changed and are no longer usable. Or perhaps none of that even matters, because the access code for the data is no longer valid anyway. We rarely consider such possibilities when we make plans, and that is why our plans often go as wrong for us as Rumpelstiltskin's did for him.

Rumpelstiltskins are everywhere. Certainly they are in politics. Here, for example, is part of a summary of the German parliament's debate on pension reform in 1972:

> The predictions in the reports on the adjustment of pensions showed that the surplus in the pension fund would rise to over 200 billion marks in the coming fifteen years. These predictions assumed solid and permanent growth in the economy, con-

stantly increasing employment, and a substantial increase in wages, amounting to seven to eight percent per year.

Simple calculations of alternative scenarios showed, of course, that these billions would melt away to nothing if these optimistic assumptions were changed. But no one gave that possibility a second thought. Both the coalition and the opposition parties looked on those 200 billion marks as if they were already in the bank and available to spend. And so they were in fact spent.[13]

Rumpelstiltskin!

Discounting the conditions under which we will carry out our actions simplifies planning but not acting. We can describe our plans much more easily—as when a general outlines a campaign as simply a matter of capturing a series of towns—but our description may conceal problems. As Carl von Clausewitz, the author of a seminal work on the conduct of war, put it, "In war everything is simple, but it's the simple things that are difficult."

Plans often fail because the planners have not factored in all the irksome little conditions, or "frictions" as Clausewitz called them, that have to be dealt with if the plan is to succeed. The plan may be simple; carrying it out is the hard part. Likewise, in the history of technology, apparently unimportant difficulties may combine to make progress much slower than the experts expect. In 1957 the psychologist, economist, and computer scientist Herbert A. Simon (later winner of the Nobel Prize in economics) predicted that within ten years a computer would be the world chess champion. He was off by at least thirty years. Only today are there chess computers capable of play on a world-class level, and these have consistently lost to the very best players.

Simon was and is an interdisciplinarian well-versed in the theory of decision making. How could he have been so wide of the mark? The story must have gone something like this: Simon knew in general terms what the requirements were for programming a successful chess computer. He had quite clear ideas and he knew that his ideas were correct. But what he didn't take into account were the many irksome little problems that usually crop up in development projects like this one. He saw the route he would have to take but not the obstacles. It is when we have the big picture firmly in mind that we are most prone to forget about details.

Not taking frictions into account is a trap that experts are particularly apt to fall into. They are often correct about the lines of development, but they tend to underestimate how long it will take to translate the possible into the actual. There are many examples of this tendency: predictions made in 1983 that 90 percent of all American households would be hooked up to a video data system by 1985, that every television set in 1985 would be able to receive three hundred channels, and that by 1985 it would be possible to do computer programming in normal language.

All these things are doable. The most difficult of them is programming computers in something approaching normal language. But 1985 has long since gone by without one of these predictions being anywhere near fulfillment.

Having shown my scorn for deconditionalized planning on the Rumpelstiltskin model, I feel I should praise it a little, for it often gives us something we sorely need, namely, optimism and courage. There are many tasks we would never dare to take on if we didn't first conceive of them in very simple terms. And if we do dare, perhaps we will succeed. "On s'engage et puis on voit!"

In science, individuals often make significant contributions outside of their fields, at least in certain phases of scientific advancement. For example, it was Erwin Schrödinger, a theoretical physicist, who propounded the belief that genes were the key components of living cells.[14] The ideas nonexperts propose need not even be strictly accurate to be valuable. The biocyberneticist Ludwig von Bertalanffy writes, "Oversimplifications, progressively corrected in subsequent development, are the most potent or indeed the only means toward conceptual mastery of nature."[15] Experts see things in much more differentiated form—that's what makes them experts—and for that very reason they may overlook other perspectives.

Deconditionalizing is not limited to single actions or single planning elements; it can determine large units of action as well. Our everyday activities are controlled automatically. We don't need to make plans for them, because we already have plans. The sequences in which we carry out most of our everyday activities are equally preestablished. These "automations" function almost like reflexes, that is, "automatically." We get dressed, make coffee, fry eggs, butter our toast, start our cars, and drive to work on our familiar route without thinking at all about the sequence of these actions. Such automations are essential in that they save us from thinking about every detail of our lives. We would never accomplish anything if much of what we do were not preprogrammed.

But we pay a price for this relief. It is possible that things would go much more simply and easily for us than they have in the past, with less friction and fewer repercussions, if we gave a little thought to the sequence of our actions.

A famous experiment in cognitive psychology demonstrated that automations, in addition to the benefits they provide, also tend to blind us to new possibilities. Abraham and Edith Luchins gave experiment participants the task of measuring certain amounts of water with the help of three pitchers.[16] Participants could fill the pitchers, pour the contents of one into another, or empty them completely. They could, for example, measure out three quarts of water with a five-quart and a two-quart pitcher by filling the five-quart pitcher and then filling the two-quart one from it, leaving exactly three quarts.

The Luchins set their experiment subjects several tasks of this kind, but all the tasks could be solved by following the same sequence of operations. If, for example, container A holds nine quarts, container B forty-two quarts, and container C six quarts, we can measure out twenty-one quarts by filling B, then filling C twice and A once. That will leave twenty-one quarts in B. This sequence, which we can abbreviate B − 2C − A, proved applicable to five tasks in a row. The sixth task read: "Container A holds twenty-three quarts, B forty-nine quarts, and C three quarts. Measure out twenty quarts."

The sequence B − 2C − A will yield the desired amount, but A − C yields the same result much more easily. Most of the experiment participants did not hit on this solution, however. (Incidentally, this is an example of how experience does not always make us smart. Experience can also make us dumb.)

The effects of "methodism"—the unthinking application of a sequence of actions we have once learned—can have a significant impact in areas other than measuring quantities of water. The motivation is the same, however: we are most inclined to deconditionalize a form of action and use it over and over again if it has proved successful for us or for others. Clausewitz, from whom I have borrowed the term *methodism*, observes:

> So long as no acceptable theory, no intelligent analysis of the conduct of war exists, routine methods will tend to take over even at the highest levels. Some of the men in command have not had the opportunities of self-improvement afforded by education nor contact with the highest levels of society and government. They cannot cope with the impractical and contradictory arguments of theorists and critics even though the commanders' common sense rejects them. Their only insights are those that they have gained by experience. For this reason, they prefer to use the means with which their experience has equipped them, even in cases that could and should be handled freely and individually. They will copy their supreme commanders' favorite device—thus automatically creating a new routine. When we find generals under Frederick the Great using the so-called oblique order of battle, generals of the French Revolution using turning movements with a much-extended front, and commanders un-

der Bonaparte attacking with a brutal rush of concentric masses, we then recognize in these repetitions a ready-made method and see that even the highest ranks are not above the influence of routine.

War, in its highest forms, is not *an infinite mass of minor events,* analogous despite their diversities, which can be controlled with greater or lesser effectiveness depending on the methods applied. War consists rather of *single, great, decisive actions,* each of which needs to be handled individually. War is not like a field of wheat, which, without regard to the individual stalk, may be mown more or less efficiently depending on the quality of the scythe; it is like a stand of mature trees in which the ax has to be used judiciously according to the characteristics and development of each individual trunk.[17]

We can summarize Clausewitz's advice thus: In many complex situations, considering a few "characteristic" features of the situation and developing an appropriate course of action in the light of them is not the essential point. Rather, the most important thing is to consider the specific, "individual" *configuration* of those features and to develop a completely individual sequence of actions appropriate to that configuration. The methodist is not able to cope with specific, individual configurations on their own terms, for he has his two or three ways of proceeding, and he uses one or the other depending on the general features of the situation as a whole. He does not take into account the individuality of the situation as it is evidenced in the specific configuration of its features.

Methodism is dangerous because a change in some minor detail that does not alter the overall picture of the situation in any appreciable way can make completely different measures necessary to achieve the same goal. Clausewitz's metaphor of felling trees is apt. To make our cuts we have to analyze carefully the tilt of the tree, the position of the neighboring trees, the shape of the tree's crown, the direction of the wind, and any unusual twists in the trunk. Any one of these features might require us to relocate our planned cuts.

Having tried-and-true methods available to us can produce optimism in planning. And that in turn may have the positive result of giving us the confidence to attempt something new. But it can also have the opposite effect. If these methods have in fact repeatedly proved effective in the past, we will be all the more inclined to overestimate their efficacy. As at Chernobyl, certain actions carried out frequently in the past, yielding only the positive consequences of time and effort saved and incurring no negative consequences, acquire the status of an (automatically applied) ritual and can contribute to catastrophe.

Methodism is likely to flourish in those situations that provide feedback on the consequences of our actions only rarely or only after a long time. In particular, if our plans apply to a field in which we rarely act, our planning gradually degenerates into the application of ritual.

Two psychologists, Jürgen Kühle and Petra Badke, conducted an interesting experiment that exposed the common propensity for deconditionalized planning.[18] Their experiment used what we might call "experimental history" or "experimental politics," by presenting the participants with the situation of Louis XVI in 1787.

In that year in prerevolutionary France, a special assembly of Notables convened by the king turned down a suggestion for tax reform put forward by Louis's controller general of finance, Charles Alexandre de Calonne; they also demanded Calonne's resignation. The principal goal of the tax reform was to take away some of the privileges the nobility and the clergy enjoyed, a measure urgently needed if France was to recover a firm financial footing.

The experiment participants were informed about the country's domestic situation, the mood of the populace, and the economic, military, and political situation following the war that had just been fought against England (when France entered the American War of Independence on the side of the American colonies). To prevent the participants from recalling points from their study of history, the experiment directors disguised the situation as taking place in ancient China circa 300 B.C.; only participants who had extensive and detailed historical knowledge would have been likely to recognize the circumstances as those of prerevolutionary France. The participants were now asked to assume the role of the king and to lay plans. They were given only forty-five minutes for this task. (In an experiment of this kind, the point is precisely not to give enough time for detailed study of all the circumstances.) After their time was up, participants were to submit a list of the measures they thought appropriate. (One participant who did *not* recognize the situation as that of Louis XVI said after reading the instructions and the description of the experiment, "I think that in a situation like this I would wind up with my head on the block." The instructions seem to have been illuminating.

Each participant's individual measures were evaluated on their degree of "elaboration." The specificity with which the participant described a goal and the measure for achieving that goal determined the "elaboration index" for that particular measure. If both descriptions were very detailed, the measure received an elaboration index of 2. If individual details were not supplied or were left vague, the elaboration index was lowered accordingly. Another aspect evaluated was a measure's degree of "conditionality," that is, the extent to which the participant determined which conditions necessary for carrying out the measure were already met and which needed to be created.

Some participants merely proposed goals and gave only cursory attention to the conditions and to the execution of the measures necessary to achieve those goals. Some participants said, for instance, "I would balance the national budget by taking money from the nobles." But these participants failed to say how exactly they would manage this feat. As the Notables' assembly had just made perfectly clear, the nobles were not prepared to give their money up willingly. Another participant simply decreed outright, "The first thing we'll do is produce an export surplus."

Many participants, however, were quite capable of producing detailed plans. The average elaboration index was under 50 percent of the maximum. It is interesting to note here that the correlation between elaboration and conditionality was quite high.[19] It also turned out that conditionality and elaboration were highly correlated with a measure's estimated *certainty of effectiveness.*[20]

Another study investigated the degree to which the individual participants coordinated their measures. Participants who showed a high elaboration index and a high degree of conditionality also devised *primary* measures (usually for carrying out tax reform and preventing financial collapse) to which they subordinated various secondary measures.

Bad participants simply compiled laundry lists of goals without providing operative principles for achieving them, and they made no distinction between primary and secondary measures. The bad participants produced a "jumble," not an "organized set," of measures. (This was a result that turned up in the Greenvale experiment as well.)

In a second experiment Kühle and Badke studied the correlation between elaboration indices for measures proposed in various political situations of this kind (including the situation of Louis XVI) and the results of the Moro planning game.[21] The correlation showed that the individuals with high elaboration indices in the political situations also tended to be successful in the Moro game. On average, these individuals produced fewer catastrophes or near catastrophes (the numbers were statistically significant) than did participants with low elaboration indices. Whatever it was that moved the good participants to plan their measures with greater elaboration and conditionality, it seems to be directly related to competence in a complex planning game.

Like Kühle and Badke, Thomas Roth used results from a planning game to identify characteristics of good problem solvers.[22] Roth studied the language that good and bad participants used while engaged in a simulation game called "Tailor Shop." The task was to manage a small plant that manufactured clothing. Using records of what the participants said as they thought out loud during the experiment, Roth studied the nature of their problem-solving language.

Roth found that the bad problem solvers tended to use unqualified expressions: *constantly, every time, all, without exception, absolutely, entirely, completely, totally, unequivocally, undeniably, without question, certainly, solely, nothing, nothing further, only, neither . . . nor, must,* and *have to.*

The good problem solvers, on the other hand, tended more toward qualified expressions: *now and then, in general, sometimes, ordinarily, often, a bit, in particular, somewhat, specifically, especially, to some degree, perhaps, conceivable, questionable, among other things, on the other hand, also, moreover, may, can,* and *be in a position to.*

It is obvious from these two lists that the good problem solvers favored expressions that take circumstances and exceptions into account, that stress main points but don't ignore subordinate ones, and that suggest possibilities. By contrast, the bad problem solvers used "absolute" concepts that do not admit of other possibilities or circumstances.

The following excerpts, which are of course too brief to reflect the whole range of results, illustrate these differences. First, a good participant:

> "Why is it that I've produced only 503 when I have the capacity to produce 550?"
>
> Experiment director: "I'm not permitted to comment on that."
>
> "There could be many reasons. How am I supposed to find out whether a worker was sick or whether a machine broke down—or what the reason was?"
>
> Experiment director: "The machines were all working."
>
> "They were working. And at full capacity? How can I find that out?"

And now a bad problem solver:

"There's no point in doing anything. I'm going on the assumption that the shirts sold before, and so they'll sell again. . . . And why aren't they selling? The fault must lie in the shirts themselves. Either there's something new on the market or the collars aren't modern or something like that. My market position is optimal. . . . I've stepped up advertising. There's nothing more I can do."

The differences are blatant: in one case, the attempt to analyze and to find reasons; in the other, dogmatism and assertions instead of analysis. We don't know whether Roth's good participants actually produced more conditionalized and elaborated measures than the bad ones, but that seems likely in view of their greater success.

Our concern here has been the treatment, consideration, or disregarding of conditions in planning. Why is it that some people plan with conditions in mind and others do not? Here we are thrown back on speculation. Nonetheless, it is worth recalling that planning that ignores conditions is easier and quicker and produces clearer concepts or action. Who will find those features desirable? Perhaps the person who feels extremely uncertain in a complex decision-making situation.

Does this mean that decisive people, with their clear, simple plans, are really nervous, wishy-washy types at heart? Sometimes maybe, but surely not always. Simple plans can be good plans, too, and sometimes decisiveness grows out of a conviction that we know what to do.

How can we distinguish one case from the other? The person whose insecurity motivates him to ignore conditions will probably know or at least suspect it. And he will consequently tend to avoid translating his plans into action because that would make their inadequacies apparent.

Before making plans in disregard of conditions, we should perhaps meditate on Kant's warning: "Making plans is often the occupation of an opulent and boastful mind, which thus obtains the reputation of a creative genius by demanding what it cannot itself supply, by censuring what it cannot improve, and by proposing what it knows not where to find."[23]

## Learn by Making Mistakes? Not Necessarily!

If we have planned, reached a decision, and implemented that decision, either by acting ourselves or by delegating the execution of our measures to others, in most cases those measures will have consequences. (Of course, measures don't always have consequences. There are some—by no means unpopular—that accomplish nothing at all.)

Studying the consequences of our measures gives us excellent opportunities for correcting our incorrect behavioral tendencies and assumptions about reality. If our measures yield unexpected consequences, there must be reasons. By analyzing those reasons, we can learn what we should do better or differently in the future. Or so one would think.

When we obtain an unexpected result, we can ask ourselves whether we began with false premises or whether we had an incorrect or incomplete or imprecise picture of reality. Then we can ask ourselves in turn why our image of reality was so far off the mark. Did we use the wrong methods for gathering information? Or did we just stop gathering information too soon? Did we form incorrect hypotheses? Or did the intransparency and complexity of the reality we were dealing with make the unexpected results unavoidable? Or were correct measures carried out incorrectly? If so, we have to keep a closer eye on the people responsible for applying them.

In any case, unexpected results should give us pause. Even negative results provide us a chance to apply correctives and in that way help us improve our future behavior. Or so one would think.

In fact, people look for and find ways to avoid confronting the negative consequences of their actions. One of these ways is "ballistic behavior."

A cannonball behaves ballistically. Once we have fired it, we have no further influence over it. The course it takes is determined entirely by the laws of physics. This is not true of a rocket, which does not behave ballistically but is subject to the control of a pilot or the remote control of an operator who can alter its flight path if it appears, that it is deviating from the designated path.

It is clear that as a general principle behavior should not be ballistic. Because our grasp of reality can only be partial, we have to be able to adjust the course of our actions after we have launched them; analyzing the consequences of our behavior is crucial for making these ex post facto adjustments.

An experiment conducted by Franz Reither demonstrates the prevalence of ballistic behavior.[24] Participants were divided into teams of five and instructed to render development aid in a fictitious region of the Sahel. Except that the local tribespeople were called Dagus, the experiment was like the Moro one. The participants could recommend the use of new kinds of fertilizer; they could introduce new breeds of cattle; they could requisition tractors, combines, and other agricultural equipment; and they could carry out other more or less useful measures. The chart on the left-hand side of figure 36 records the frequency with which participants controlled these measures. "Controlling" in this case means asking questions such as "What results have the newly introduced fertilizers produced?"

In the first five years of the teams' activity, the average control figure was 30 percent; that is, out of a hundred decisions the participants made, they later inquired about the results of thirty of those decisions. That isn't very many. But, as we also see in figure 36, things changed during the course of the experiment. In the second five-year period, control rose to more than 50 percent.

The desire to check on efficacy seems to increase but never assumes major proportions. That is odd because we would expect that

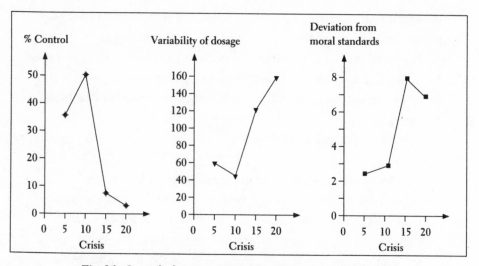

Fig. 36. Control of measures, variability of dosage, and deviations
from moral standards in the Reither experiment

rational people faced with a system they cannot fully understand would seize every chance to learn more about it and would therefore behave "nonballistically." For the most part, however, the experiment participants did not do that. They shot off their decisions like cannon-balls and gave hardly a second thought to where those cannonballs might land.

Strange, we think. But an explanation is readily available. If we never look at the consequences of our behavior, we can always maintain the illusion of our competence. If we make a decision to correct a deficiency and then never check on the consequences of that decision, we can believe that the deficiency has been corrected. We can turn to new problems. Ballistic behavior has the great advantage of relieving us of all accountability.

The less clear a situation is, the more likely we are to prop up our illusion of competence with ballistic behavior. Such behavior reduces our sense of confusion and increases our faith in our own capabilities. And that isn't necessarily bad, is it?

There is more to be said about Reither's Dagu experiment. After the tenth year, a crisis occurred: a neighboring state laid claim to 30 percent of the Dagus' territory and, in an act of brutal aggression, simply occupied it. The Dagus retained enough land for their needs, however, and this fact was communicated to participants. In other words, there was no real need to respond to the occupation with extraordinary measures.

As Reither describes the response:

Decisions to purchase weapons and to provide military training to a population with no previous military experience were reached with relative unanimity. To raise funds for these additional expenses, the participants then decided they needed a major increase in yields both from field crops and from cattle raising, and to this end they radically increased the use of fertilizers and pesticides and drew more heavily on the groundwater supply. Conscription of some of the male population meant a reduction in the workforce, and participants tried to compensate by demanding more work from the remaining workers and especially from women and children. And this measure was frequently accompanied by food rationing.

People do not, of course, take a crisis of this kind lightly, and it is easy to see why the experiment participants felt they needed to change their behavior radically. But because changing their behavior forced them to abandon strategies they had become comfortable with, their sense of their competence suffered. As we can see from figure 36, the situation produced a massive increase in ballistic behavior. In the years 11–15 and 16–20, 10 percent or less of all decisions were subjected to follow-up control. The vast majority of decisions were ballistic, lending support to our idea that ballistic behavior is used to bolster the illusion of one's competence.

The increase in ballistic tendencies was not the only effect of the crisis after year 10. The middle chart in figure 36 illustrates another interesting effect: the degree of variability in the "dosage" of measures. "Dosage" here means the intensity with which a measure is carried out. If, for example, we apply three tons of fertilizer to an acre of land, we are applying a higher dosage of that measure than if we use two tons per acre. If we requisition the drilling of thirty wells, that is a higher dosage than a requisition for only ten wells.

Reither measured the variability of dosages.[25] If all the participants apply just about the same dosage, the variability is low. If some participants apply very high dosages and others very low ones, the variability is high. The middle chart shows the course of the variability coefficient Reither obtained over the four five-year blocks of the experiment. The measures the participants took before the crisis show only moderate variability, but after the crisis there is a marked, and statistically significant, change: the participants resorted either to stronger means or to weak dosages indicative of resignation. These results also support the idea that activity may foster an illusion of competence. By intervening massively, a person demonstrates his competence, his ability to take the situation in hand—he demonstrates it to himself at least. Conversely, someone who feels obliged to demonstrate to himself or to others a competence he believes he does not possess may well yield to resignation.

A third consequence of the crisis in the Reither experiment may be of even greater interest than the results I have discussed so far. The chart on the right-hand side of figure 36 shows this effect.

After the experiment, Reither asked his participants to look once more at the various measures they had taken, but this time he asked them to rank the measures in terms of how much they deviated from the participants' moral and ethical standards. In the first two sessions, the average number of deviations was relatively small. But after the crisis, the deviations were quite large, averaging between 6 and 7 points. Not only did the crisis and the loss of competence accompanying it cause participants to act more ballistically and to raise the dosages of their measures, but the participants began to act on the principle that "ends justify means" and paid less attention to overarching moral standards. We find here, in short, a drift toward cynicism and the erosion of moral standards. This is, of course, nothing new. But the fact that a drift of this kind emerges as a statistically significant effect—that is, as a general tendency in what is, after all, an altogether unthreatening simulated situation—certainly gives us something to think about.

Ballistic behavior is not the only way we can avoid confronting the negative consequences of our own actions. If there is no other escape and we are forced to recognize them, we can always resort to what psychologists neatly term "external attribution." We can always say, "I had the best of intentions, but circumstances prevented me from achieving what I wanted." "Circumstances" can, of course, always be found, especially those "forces of evil" that sabotage and thwart our finest efforts with malevolent and underhanded tactics.

Yet another form of interpreting away one's failings is to invert goals. I mentioned this tactic before: we can make "good" out of "bad" and assign to the famine we have created in Tanaland an important role in improving the population structure.

Finally, we can protect and maintain our competence by an "immunizing marginal conditionalizing" of our measures.[26] As a rule, measure A produces effect B, we can reason. But under certain limited conditions that are unfortunately prevailing just at this moment, measure A produces other effects.

One participant in the storeroom experiment was convinced that odd numbers raised the temperature in the storeroom and that even numbers lowered it. But on one occasion the temperature dropped when the regulator was set at an odd number. Did that prove to our participant that he was wrong? Not at all. You see, if you set the regulator at 100 immediately before you set it on an odd number, the general rule doesn't apply. In that special case, the odd numbers have a different effect.

In the storeroom experiment, the feedback and the confrontations participants had with the consequences of their actions came too frequently for anyone to be able to protect incorrect hypotheses for very long by conditionalizing on the basis of local and therefore irrelevant circumstances. In other situations, however, such conditionalizing may be possible. If feedback on the consequences of our actions comes rarely and is of a kind that can be easily ignored, immunizing marginal conditionalizing is a marvelous method for dispelling all doubts about our competence.

# Seven

## So Now What Do We Do?

We have become acquainted with many inadequacies of human thought in dealing with complex systems. We have seen people fail to formulate their goals in concrete terms, to recognize when their goals contradict one another, and to set clear priorities. We have also seen them badly mishandle temporal developments. Above all, we have seen people fail to correct their errors. It may be possible to correct these failings by rote, but a more useful exercise will be to determine the main psychological reasons for these inadequacies so that we can attack them at their roots.[1]

The first reason for many failings is simply the slowness of human thinking. I deliberately say "thinking" here, in contrast to our unconscious information processing. We humans are quite deft at many tasks. The speed with which an average driver can process, and respond correctly to, a wide variety of information in heavy traffic is staggering, and it evokes considerable respect from anyone who has ever tried to develop, even conceptually, an artificial system as good. What artificial systems have been able to achieve in this respect is not impressive, and one can only smile at the difficulty they have in sifting out from a complex and variegated environment the right information they need to guide correct actions. (Development of such systems has by no means reached a dead end, however, and I certainly do not believe that they are by nature incapable of identifying complex forms and filtering information.)

On the other hand, if a computer could laugh, it would surely find utterly ridiculous the amount of time it takes us to divide 341,573 by 13.61. We can practice such tasks and increase our speed at them. We can memorize the prime numbers between 1 and 10,000 and thereby incorporate many shortcuts into such thought processes. But such efforts wouldn't change the fact that our conscious thought, the very "tool" we need to deal with unknown realities, functions rather slowly and is not capable of processing many different pieces of information at the same time.

It is no wonder then that our slowness obliges us to take shortcuts and prompts us to use our scarce resources as efficiently as we can. This need to economize—to save time and effort—underlies many of the failings in our thought processes that I have examined here. Let's look at some individual examples.

The motto "first things first" may well explain why, when confronted with a task, we immediately begin planning our actions and gathering information instead of formulating our goals in concrete terms, balancing contradictory partial ones, and prioritizing them. We've got a problem, so let's get to it and not waste a lot of time developing clarity about it.

If, instead of clarifying the complex interrelationships among variables in a system, we select one variable as central, we are economizing in two respects: first, we dispense with a great deal of additional analytical work; second, we save time later in gathering information and in planning, for if one variable is central to the entire problem at hand, we need information only about that variable. If everything else depends on it anyway, we don't have to worry about the status of other variables. We can also focus our planning on the one core variable. Reductions of this type are hard to beat. They allow us to make the most economical use of that precious resource we call "thought."

When we deal with a complex system of variables by setting up rules, we are again economizing in two respects: first, we eliminate the need to sort through the confusing variety of circumstances under which a certain action was successful; second, we streamline our planning by using only a few general rules rather than many rules that are only locally applicable and for which we must determine, case by case, whether the conditions necessary for their successful application exist. "Strategic thinking," which Clausewitz stressed in his tree-felling metaphor, demands a far greater expenditure of mental energy than does thinking that uses a single principle the way a mower wields a scythe.

If, in dealing with temporal configurations, we extrapolate linearly, we spare ourselves much of the thought that goes into the complex and tiresome observations and analyses needed to understand the specific laws governing any particular temporal process.

Planning without taking side effects and long-term repercussions into account is also far more economical than analyzing those possibilities in advance.

"Methodism"—seeing new situations in terms of old, established patterns of action that need only be set rolling—is far more economical than considering in each individual case what the specific local conditions demand in the way of response.

"Ballistic decisions" that allow us to ignore the consequences of our actions save us a lot of thought, too. They let us avoid time-consuming (and unsettling) reflections about how we could perhaps have done things better.

In short, our tendency to economize, which prompts us to omit certain steps in the thought process or to simplify them as much as possible, seems to play a major role when we deal with complex systems.

A second reason for many inadequacies and errors in human

thought lies outside the realm of cognitive processes. Preserving a positive view of one's competence contributes significantly to shaping the direction and course of our thought processes.

When we are asked to act, we do so only if we feel at least minimally
competent to do what is asked of us. We need to feel that our actions
will ultimately be successful. Without some expectation of success, we
are unlikely to act at all and will rather resign ourselves to letting fate
take its course. We often redirect our thinking from our actual goals to
the goal of preserving a sense of our competence. This act of self-
protection is essential to maintaining a minimum capacity to act.

Many shortcuts and omissions in thinking that we may attribute to
an effort to economize can also be interpreted as self-protection. If we
develop a reductive hypothesis and see everything as dependent on a
central variable, we not only make things easier, we also derive the
reassuring feeling that we have things under control. Without this sim-
plification, we might find ourselves afloat on a sea of data and interrelationships that are far from easy to analyze, and being at sea is not a
pleasant feeling. Forming simple hypotheses and limiting the search for
information shortens the thought process and allows a feeling of competence.

Our tendency to pursue planning, information-gathering, and struc-
turing processes that go on interminably can also reflect a need for self-
protection. If excessive planning and information gathering keep us
from making contact with reality, reality will have no opportunity to tell
us that our measures aren't working or are all wrong.

"Methodism" may also arise from self-protective tendencies. Rather
than think about the specific demands of a specific situation, as Clausewitz's tree-felling metaphor suggests we should, and rather than discover
that the schemes for action we have ready to hand are not applicable,
we prefer to assume that the new problem is of an old, familiar type that
we have solved frequently in the past. This assumption makes us feel secure—we see that we can cope with the situation. Then if we actually
have to do what we've planned, we can avoid confronting any errors—

or the simple fact that our action had no effect at all—by acting ballistically. We will simply refuse to look at the consequences of our actions.

Another proven means of protecting our sense of competence is to solve only those problems we know we can solve. If we solve the problems we can and avoid the ones we cannot, we reinforce our sense of competence.

A third reason we have difficulty dealing with complex and time-dependent systems is the relatively slow speed with which the storage system of the human memory can absorb new material. Human memory may have a very large capacity, but its "inflow capacity" is rather small. What we perceive at any given moment may be rich in content, colorful, and clear in its contours. The moment we close our eyes, however, a great deal of that richness instantly disappears—unclear and pale outlines remain—and the farther back into the past we go, the poorer in information our memory records of events become.

This fading of received information may have its function. It may serve to provide us with those abstract schemes we need for forming "classes of equivalence" by shielding us from a superfluity of information. Surely, however, it has its disadvantages. The difficulty we human beings have with temporal configurations is an important example. That such processes present an excess of information may partly account for these difficulties, but they may also result partly from our forgetfulness, our loss of information. If we cannot form a picture of a temporal configuration, we cannot adjust our thinking and actions to take that temporal pattern into account. This inability explains the ad hoc behavior of some participants in the storeroom experiment and the obsession with the status quo of some participants in the moth experiment. They couldn't think about everything that had happened earlier, simply because that information was no longer present in their memory.

A fourth psychological mechanism seems responsible not so much for failings as for omissions in our thought processes. We don't think about problems we don't have. Why, indeed, should we? In solving problems that involve complex dynamic realities, however, we must think about problems we may not have at the moment but that may emerge as side effects of our actions.

We don't neglect the "implicit" problems of a situation because thinking about the possible side effects of the measures we are planning would overburden us terribly. Rather, we neglect them because we don't have those problems at the moment and therefore are not suffering from their ill effects. In short, we are captives of the moment.

The slowness of our thinking and the small amount of information we can process at any one time, our tendency to protect our sense of our competence, the limited inflow capacity of our memory, and our tendency to focus only on immediately pressing problems—these are the simple causes of the mistakes we make in dealing with complex systems. But because they are comprehensible stumbling blocks, we should be able to find ways to avoid them most of the time. In what follows, we will consider some possibilities for improvement.

Let's return once again to the Moros. Figure 37 shows the average number of cattle, the vegetated area, the groundwater, the capital, and the yield of the millet crop at the end of the "reigns" of two groups of fifteen participants each: "practitioners" in the field of planning and decision making and "laymen" in that field.

We set both groups to the task of managing the Moro system. The practitioner group was made up of managers from large industrial and commercial firms; the layman group was made up of students. As figure 37 makes clear, in terms of almost all the criteria the practitioners left the land of the Moros in a far better state than the laymen did. In some variables the groups showed no differences. The millet yield for both groups showed little difference on average, and the same is true of the groundwater supply. For the critical variables of capital, cattle stock, and vegetated area, however, the differences are significant indeed.

Why is this so? Our practitioners all held senior management positions in business and industry. They were much older than our student participants, and they consequently had much more professional and life experience behind them. We did not run intelligence tests on our

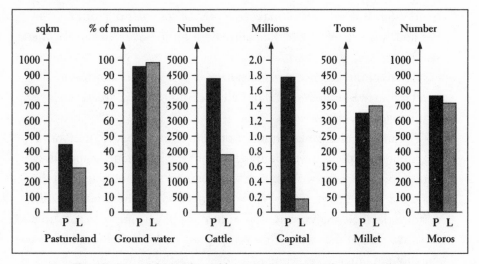

**Fig. 37.** Status of critical variables in Moro territory after twenty years of governorship by "practitioners" (P) and "laymen" (L)

participants from business and industry, and so we can only guess as to whether our two groups were comparable as far as intelligence goes. Our assumption is that our students did not fall behind the managers in this respect. Indeed, it was our impression that the students were somewhat quicker and better able to note and retain information than the older participants. We would expect to find such differences between older and younger participants. Nevertheless, our older participants, with their greater experience in planning and decision making, performed considerably better. Other studies have yielded very similar results. For example, in an experiment focusing on economic management, business-school professors performed better than business-school students. In this case, however, the professors' command of the field may have played a role.[2]

None of the participants in either of our groups could draw on previous management experience in the Sahel region, but, given current interests among students, we can probably assume that our laymen had a greater interest in ecology and in the Third World than the practitioners did. But we cannot make any precise statements about the values and interests of our participants. Both groups seemed to take the same interest in the experiment.

If neither intelligence nor specialized experience nor motivation differed in the two groups, what did? What accounts for the greater success of the practitioners?

I think that the explanation is "operative intelligence," the knowledge that individuals have about the use of their intellectual capabilities and skills. In dealing with complex problems we cannot handle in the same way all the different situations we encounter. Sometimes we must perform detailed analyses; at other times it is better simply to size up a situation. Sometimes we need a comprehensive but rough outline of a situation; at other times we may have to give close attention to details. Sometimes we need to define our goals very clearly and analyze carefully, before we act, exactly what it is we want to achieve; at other times it is better simply to go to work and muddle through. Sometimes we need to think more "holistically," more in pictures, at other times more analytically. Sometimes we need to sit back and see what develops; at other times we have to move very quickly.

The remarkable fact is that we are capable of acting in all these different ways: "We do not need to reorganize our brains; all we need to do is make better use of their possibilities."[3]

Everything at its proper time and with proper attention to existing conditions. There is no universally applicable rule, no magic wand, that we can apply to every situation and to all the structures we find in the real world. Our job is to think of, and then do, the right things at the right times and in the right way. There may be rules for accomplishing this, but the rules are local—they are to a large extent dictated by specific circumstances. And that means in turn that there are a great many rules.

I think that the differences between the expert and the layman can be found here. We all know the basic rules of thumb. "Look before you leap." "Be clear about your goals." "Gather as much information as you can before you act." "Learn from your mistakes." "Don't act in anger."

"Ask for advice." Who would not agree to their usefulness? The troublesome thing about them is that they don't always apply. There are situations in which it is better to act than to think. Sometimes we should cut short our information gathering. And so on.

Our practitioners not only knew these rules but applied the right rules at the right times.

Our everyday language distinguishes many aspects of intellectual capability, some innate, others acquired. Geniuses are geniuses by birth, whereas the wise gain their wisdom through experience. And it seems to me that the ability to deal with problems *in the most appropriate way* is the hallmark of wisdom rather than of genius.[4]

If that is so, then it must be possible both to teach and to learn how to think in complex situations. Some of the results presented in this book show that people can respond to circumstances and learn to deal with specific areas of reality. The behavior of participant 04 mlg in the moth experiment is one illustration. As we can see from figure 34, this participant performed poorly at first but then learned from his mistakes. What this one person was able to do should be possible on a much larger scale.

How can we teach people to deal effectively with uncertainty and complexity? Here, too, finding the right strategy is essential. There is probably no cut-and-dried method for teaching people how to manage complex, uncertain, and dynamic realities, because such realities, by their nature, do not present themselves in cut-and-dried form.

Rather simple methods can, however, improve our ability to think. Figure 38 shows the result of an experiment in which participants were asked to solve a number of fairly demanding problems involving an array of lights.[5] Twelve colored lights were arranged into three components, each containing a blue, a green, a red, and a yellow light. In any "state" of the array, one light of each component would be illuminated. Each problem required the array to be changed from a given "initial state," say "red-green-red," to a particular "desired state," say "blue-yellow-green."

The participants used a keyboard that controlled the components of

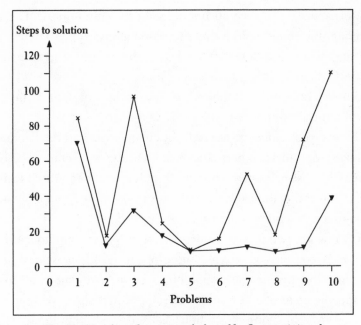

**Fig. 38.** Number of steps it took the self-reflection (▼) and
control (+) groups to solve a problem

the array. Pressing a certain "operator" key, for instance, would make
the third component rotate through all its colors. If the third component happened to be red, pressing the key would change it to green.
Pressing the key again would change green to yellow, still again would
change yellow to blue, still again would change blue back to the original red. There was also an "exchange" key that switched the colors in
the array. Pressing this key would change "red-yellow-green" to "green-
yellow-red." And finally, there were keys with very complicated effects
that depended on the existence of certain conditions. For example, if
component 1 was red and component 2 green, pressing the key would
make component 3 yellow. But if components 1 and 2 were both green,
pressing the key would make component 3 blue. The participants did
not know before they began the first problem what effects the keys
produced.

The participants were divided into two groups—an experimental group and a control group. Both were asked to solve ten problems. The members of the control group were asked after each problem to record their hypotheses about the effects of the keys. The members of the experimental group were asked simply to think about their own thought processes. They were asked, in other words, to reflect on their experiences in thinking about solving each problem.

As we see from figure 38, the request for reflection had a major effect. The experimental group performed much better than the control group. Thinking about our own thinking—without any kind of instruction—can make us better problem solvers.

We should note, however, that the array of lights is a relatively straightforward system. In more complicated situations, such unstructured reflection on our own thinking may be disruptive, make us uncertain, and so produce negative results. The Greenvale experiment shows that instruction, too, can fail to achieve its purpose when the situation is complex.

We divided the Greenvale participants into three groups: a control group, a strategy group, and a tactics group. The strategy and tactics groups received instruction in some fairly complicated procedures for dealing with complex systems. The strategy group was introduced to concepts like "system," "positive feedback," "negative feedback," and "critical variable" and to the benefits of formulating goals, determining and, if necessary, changing priorities, and so forth. The tactics group was taught a particular procedure for decision making, namely, "Zangemeister efficiency analysis."[6]

After the experiment, conducted over several weeks, the participants were asked to evaluate the training they had received; figure 39 shows the results. The members of the strategy and tactics groups all agreed that the training had been "moderately" helpful to them. The members of the control group, who had received training in some nebulous, ill-defined "creative thinking," felt that their training had been of very little use to them. The differences in the evaluations are statistically significant. If we look at the participants' actual performance as well as at their evaluations of the help they thought they got from their training, however, we find that the three groups did not differ at all in their performance.

Why did the participants who had been "treated" with certain procedures think this essentially useless training had been somewhat helpful? The training gave them what I would call "verbal intelligence" in the field of solving complex problems. Equipped with lots of shiny new concepts, they were able to *talk* about their thinking, their actions, and the problems they were facing. This gain in eloquence left no mark at all on their performance, however. Other investigators report a similar gap between verbal intelligence and perfor-

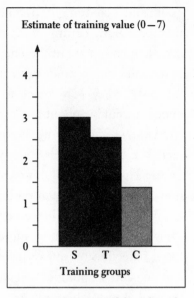

Fig. 39. Estimates of the value of the training the strategic, tactical, and control groups (S, T, and C) received in the Greenvale experiment

mance intelligence and distinguish between "explicit" and "implicit" knowledge.[7] The ability to talk about something does not necessarily reflect an ability to deal with it in reality.

Apparently, pure instruction is not necessarily of value, even when the need to apply it in the real world is immediately at hand, and it is no substitute for experience. So what can we do? The answer to this question—as to so many others we have asked—involves a balance. We should reflect on our own thinking, but with some guidance.

Although the reader may think me obsessed by computers when I suggest that these research tools so often employed in psychological studies are also valuable as instructional tools, computers can indeed provide opportunities for reflection. Critical and complicated situations are not always available for study, and in the real world, the consequences of our mistakes are slow in developing and may occur far from where we took action. After a long delay or at a great distance, we may not even recognize them as results of our behavior. We therefore have few opportunities to learn from our mistakes. A planning and decision-making scenario simulated on a computer may be less complex than one in the real world, but it has the great advantage of letting us run our experiments on fast-forward and so of bringing us face to face with our mistakes.

Thus, simulated scenarios are an excellent teaching device. But it probably profits no one if we simply turn our pupils loose on these scenarios. Action alone is of little value. What makes more sense is to assemble a battery of different scenarios that expose our participants to a "symphony of demands" posed by various systems. We should also have experts observe participants as they plan and act. These observers could pinpoint cognitive errors, and identify their psychological determinants. In carefully prepared follow-up sessions, the participants could be shown the kinds of mistakes they made and the probable causes.

What can we learn from such training?

We can learn that it is essential to state goals clearly. We all know we should do that, but we rarely encounter the necessity.

We can learn that we cannot always realize all our goals at once, because different goals may contradict one another. We must often compromise between different goals.

We can learn that we have to establish priorities but that we cannot cling to the same priorities forever. We may have to change them.

We can learn that in dealing with a given configuration we should form a model of the system. We must anticipate side effects and long-term repercussions and not just let them roll over us.

We can learn how to adapt information gathering to the needs of the task at hand, neither going into excessive detail nor stopping too short.

We can learn the consequences of excessive abstraction.

We can learn the consequences of hastily ascribing all events in a certain field to one central cause.

We can learn when to continue gathering information and when to stop.

We can learn that we tend toward "horizontal" or "vertical" evasion and that the tendency can be controlled.

We can learn that we sometimes act simply because we want to prove to ourselves that we *can* act.

We can learn the dangers of knee-jerk "methodism."

We can learn that it is essential to analyze our errors and draw conclusions from them for reorganizing our thinking and behavior.

What matters is not, I think, development of exotic mental capabilities. What matters is not full utilization of the neglected right half of the brain, not the liberation of some mysterious creative potential, and not the mobilization of that fallow 90 percent of our mental capacity. There is only one thing that does in fact matter, and that is the development of our common sense.

Everything depends, of course, on just how we employ this common sense. Temporal configurations, for example, often seem beyond common sense. As a rule we do not give adequate attention to the characteristics of processes that unroll over time. What we did yesterday is lost in the obscurity of the past, and what we ought to do tomorrow is in utter darkness. We human beings are creatures of the present. But in the world of today we must learn to think in temporal configurations. We must learn that there is a lag time between the execution of a measure and its effect. We must learn to recognize "shapes" in time. We must learn that events have not only their immediate, visible effects but longterm repercussions as well.

We also must learn to think in terms of systems. We must learn that in complex systems we cannot do only one thing. Whether we want it to or not, any step we take will affect many other things. We must learn to cope with side effects. We must understand that the effects of our decisions may turn up in places we never expected to see them surface.

Can we possibly learn all this?

We can't, in the real world, where expanses of time and space hide our mistakes from us. Hence my plea for simulations. Time passes quickly in a computer, and distance doesn't exist. A simulation can make apparent the consequences of our decisions and plans. And in this way we can develop a greater sensitivity to reality.

Mistakes are essential to cognition. But when we are dealing with real complex systems, it is hard to pinpoint our errors. In the real world, crises are (fortunately!) uncommon; there is rarely occasion for an individual to bring experience gained in one crisis to another of the same kind. As a result, mistakes made in handling demanding situations tend to teach us little of value. Simulations, by contrast, can place people in the same kind of crisis again and again to hone their sensibilities to the specific features of such situations.

My purpose is not to promote a specific mode of thought. I hope I've made clear that what is often called "systemic thinking" cannot be regarded as a single unit, as a particular, isolated capability. It is instead a bundle of capabilities, and at the heart of it is the ability to apply our normal thought processes, our common sense, to the circumstances of a given situation. The circumstances are always different! At one time this component will be crucial, at another time that component. But we can learn to deal with different situations that place different demands on us. And we can teach this skill, too—by putting people into one situation, then into another, and discussing with them their behavior and, most important, their mistakes. The real world gives us no chance to do this.

We have the opportunity today to undertake this kind of learning and teaching. Make-believe has always been an important way to prepare ourselves for the real thing. We should use this method in a focused manner. We now have far better tools for this purpose than we have ever had before. We should take advantage of them.

Is that a frivolous idea? Playing games in dead earnest? Anyone who thinks play is nothing but play and dead earnest nothing but dead earnest hasn't understood either one.

# Notes

## Introduction

1. T. Kleyn and J. Jozefowicz, "Wasteland Created by Human Hands"; reviewed in *Hamburg Evening News*, 28–29 December 1985.
2. R. Riedl, *Über die Biologie des Ursachen-Denkens: Ein evolutionistischer, systemtheoretischer Versuch*, Mannheimer Forum (Boehringer), 1978–79.
3. G. Vollmer, *Wissenschaft mit Steinzeitgehirnen?* Mannheimer Forum (Boehringer), 1986–87.
4. E. De Bono, *New Think: The Use of Lateral Thinking in the Generation of New Ideas* (New York: Basic, 1968); F. Capra, *The Turning Point: Science, Society and the Rising Culture* (New York: Simon & Schuster, 1982).

## 1. Some Examples

1. A disaster of this sort could be called Malthusian after the British economist Thomas Robert Malthus (1766–1834), who thought that all humanity was headed for just such a catastrophe. His view is not accepted today, but local in-

201

stances of this kind of development remain possible. See, for example, H. Birg, "Die demographische Zeitwende," *Spektrum der Wissenschaft*, part 1 (1989): 40–49.

2. The reader may wonder how a computer game can simulate psychological factors such as citizen satisfaction. We did this—very precisely, by population segments—by assigning numerical values to various aspects of the citizens' lives, such as standard of living, housing, prospects in the labor market, degree of crime (in Greenvale there was next to none), availability of leisure activities, and so forth (1 = "very good," for example, and 0 = "very bad"). We added up these numbers, taking into account the importance of each category, and then assigned the label "satisfaction" to the total. (The actual procedure was somewhat more complicated, but the point is that all the tallying and calculating produced a figure that our participants accepted as accurately reflecting the satisfaction of Greenvale's citizenry.

3. T. Stäudel, *Problemlösen, Emotionen und Kompetenz* (Regensburg: Roderer, 1987).

4. For those who are interested in exact figures, the results show a low positive correlation in the range of a product-moment correlation of 0.1. We assume that this correlation would be statistically significant if we were dealing with a large number of people. This correlation is so low, however, that it is of no value for prognosis or diagnosis.

5. J. T. Reason, "The Chernobyl Errors," *Bulletin of the British Psychological Society* 40 (1987): 201–06; see also David Mosey, *Reactor Accidents* (Surrey: Nuclear Engineering International Special Publications, 1990), 81–98.

6. I. Janis, *The Victims of Groupthink: A Psychological Study of Foreign-Policy Decisions and Fiascos* (Boston: Houghton Mifflin, 1972).


## 2. The Demands


1. H. Thiele, *Zur Definition von Kompliziertheitsmassen für endliche Objekte, Organismische Informationsverarbeitung: Zeichenerkennung, Begriffsbildung, Problemlösen*, ed. F. Klix (Berlin: Akademie-Verlag, 1974).

# 3. Setting Goals

1. R. Oesterreich, *Handlungsregulation und Kontrolle* (Munich: Urban & Schwarzenberg, 1981).
2. M. Csikszentmihalyi, *Flow: The Psychology of Optimal Experience* (New York: HarperCollins, 1991).
3. M. Horkheimer, "Zum Begriff der Verantwortung," *Die Verantwortung der Universität*, Weltbild und Erziehung (1954): 86.
4. C. E. Lindblom, "The Science of 'Muddling Through,' " *Readings in Managerial Psychology*, ed. R. S. Levit and L. L. Pondy (1964).
5. K. Popper, *The Open Society and Its Enemies*, vol. 2 (Princeton: Princeton University Press, 1966).
6. Tim Tisdale, one of those conducting the experiment, reported this incident to me.

# 4. Information and Models

1. See, among others, H. Gruhl, *Ein Planet wird geplündert* (Frankfurt/Main: Fischer, 1975); F. Vester, *Ballungsgebiete in der Krise* (Stuttgart: Deutsche Verlagsanstalt, 1976); F. Vester, *Neuland des Denkens* (Stuttgart: Deutsche Verlagsanstalt, 1980); and Hancock, *Lords of Poverty* (New York: Atlantic Monthly Press, 1989).
2. I have borrowed this usage of the term *critical* from Vester (*Ballungsgebiete*, 61), who speaks of "critical elements" rather than "critical variables."
3. The fact that seals in the least polluted parts of the North Sea were the first to die had little effect on the public's hypothesizing. See H. Schuh, "Der Rummel um die Robben," *Die Zeit*, 8 July 1988.
4. See V. Gadenne and M. Oswald, *Entstehung und Veränderung von Bestätigungstendenzen beim Testen von Hypothesen* (Mannheim: Fakultät für Sozialwissenschaften, 1986).
5. F. von Schmerfeld, ed., *Graf von Moltke: Ausgewählte Werke*, vol. 1 (Berlin, 1925), 241–42. I am grateful to Hans Hinterhuber of the University of Innsbruck for calling my attention to Moltke's writings on planning and strategic thinking.
6. C. Duffy, *Frederick the Great: A Military Life* (New York: Atheneum, 1986).
7. R. von der Weth, "Die Rolle der Zielbildung bei der Organisation des Handelns" (dissertation, Faculty of Pedagogy, Philosophy, and Psychology, University of Bamberg, 1989).

8. J. Goebbels, *Final Entries, 1945: The Diaries of Joseph Goebbels*, ed. H. Trevor-Roper (New York: Putnam, 1978).

## 5. Time Sequences

1. Given a quantity $k_0$ growing at a rate p, the compound-interest formula gives its value at stage n of its growth:

$$k_n = k_0 \times (1 + p/100)^n$$

This is the formula we learn in school to calculate that $10 invested in the year 1800 at 6 percent interest would have yielded $860,028 by 1995. Who among us hasn't lamented the shortsightedness of his ancestors but also doubted the reliability of the formula?

2. A. Bürkle, "Eine Untersuchung über die Fähigkeit, exponentielle Entwicklungen zu schätzen," term paper, Department of Psychology, University of Giessen, 1979.

3. *Frankfurter Allgemeine Zeitung*, 14 September 1985.

4. For example, to calculate the number of cases after ten years, substitute the values $k_0 = 262$, $p = 130$, and $n = 10$ into the equation in note 1, above. Then $k_{10}$ = 262 × (1 + 130/100) = 1,085,374.6.

5. *Die Zeit*, 15 October 1985, 13 November 1985.

6. *Fränkischer Tag* (Bamberg), 13 December 1985.

7. *Fränkischer Tag*, 2 December 1988; *Abendzeitung* (Munich), 1 December 1988.

8. This relationship between rate of growth (p) and doubling time (dt) can be made mathematically precise using the following formula:

$$dt = \ln (2)/[\ln (1 + p/100)],$$

where ln denotes the natural logarithm. The larger a number, the larger its natural logarithm; thus, as p increases, we are dividing by a larger number to get the value of dt, which means that dt decreases. Conversely, as p decreases, dt increases.

9. For HIV infections in Germany, we have only the figures that laboratories have been required to furnish since the fall of 1987. Because we do not know what the relationship is between the number of infections reported by the labs and the total number of new infections, all we can say on the basis of the lab reports is that there are *at least* as many HIV-infected persons in West Germany as the lab reports indicate. The actual number remains unknown.

10. The increase in the number of those infected can be calculated with the following formula:

    newinfe = oldinfe/(pop − 1) × (pop − oldinfe) × parte × probinfe.

    *newinfe*: the number of those newly infected
    *oldinfe*: the number of those already infected
    *pop*: the size of the population
    *parte* (partner change): the relative frequency of partner change in the population. A parte of .2 means that each month 20 percent of the population look for and find new sexual partners.
    *probinfe*: the probability that someone who lives with an infected person will become infected too.

    If we begin with one infected person, then after one month there will be $1 + (1/999) \times 999 \times .2 \times .8 = 1.16$ infected persons; after two months, $1.16 + (1.16/999) \times 998.84 \times .2 \times .8 = 1.3455$ infected persons; after three months, 1.5607 infected persons, and so on. (Naturally there will be no fractions of infected persons. The decimal figures are best understood as estimated averages.)

    With this formula we can determine for any point in time the number of new infections. (This formula is based, of course, on a number of assumptions that not everyone may consider correct. For example, it is assumed here that the changes of partners occur completely at random and that there are no subpopulations who may act on certain preferences.)

    If the population remains completely unchanged and if all the other parameters remain the same, the formula above will yield the growth over time in the number of infections according to the logistic equation

$$y = 1/(1 - \exp [-a \times (t_h - t)]),$$

    where a indicates the "steepness" of increase and $t_h$ the temporal halfway point of the increase—that is, in the case of an epidemic, the point in time at which half the population is infected. I will not go into the details of the relationship between this formula and the preceding one.

11. J. J. Gonzales and M. G. Koch, "On the Role of Transients for the Prognostic Analysis of AIDS and Anciennity Distribution of AIDS Patients," *AIDS-Forschung* 11 (1986): 621–30; "On the Role of Transients (Biasing Transitional Effects) for the Prognostic Analysis of the AIDS Epidemic," *American Journal of Epidemiology* 126 (1987): 985–1005.

12. Recall that the formulas in note 10, together with our initial assumptions, will

tell us the number of HIV-infected people at any given time. We now want to know when these people will fall ill.

Mathematically, the simulation is based on the assumption that, at any given time (t) after infection, the probability (p) of illness is:

$$p = 1 - \exp\left(-r \times [t - t_i]\right),$$

where $t_i$ represents the time of infection. (Although I don't discuss in detail here why I use this particular formula, the reader should look at it simply as a precise hypothesis on how the probability of an infected person's falling ill with AIDS increases over time.)

The choice of a value for r reflects our hypothesis about the average length of time it takes until an infected person falls ill. If we take the value 0.00015 for r, we get an average incubation period of 96 months. If $t - t_i =$ 80, for example—that is, if the population has been infected for 80 months— the formula above yields a value of 0.01193. If we multiply the population that is still healthy times this factor, we get the number of transitions from "infected" to "ill." (In our example, about 625 people are still healthy at month 80. So 615 × 0.01193 = 7.34, which is approximately equal to the figure of 0.75 percent of 1,000 given above for the number of transitions from "healthy" to "ill.")

We can now obtain the results in figure 24.

Starting from 46 infected individuals, the formula above and the formulas in note 10 work together to tell us when these 46 infected people will fall ill, how many new infections next appear, when the newly infected individuals will fall ill, and so on, generating the solid line that runs through the data points.

13. That our choice of initial parameters is arbitrary does not mean that one other result of this study, that is, the number of individuals infected, is arbitrary. We made a second simulation in which we assumed that, from May 1986 on, radical changes in the behavior of the at-risk population lowered the infectiousness rate from 53 to 1 percent, with this drop occurring gradually. In June 1986, for example, the infectiousness rate dropped to 52.57 percent, in July 1986 to 52.14 percent, and so on. These changes were clearly reflected by a rapid drop in the growth rate of infected individuals. By the end of 1992 their number represented only 7.6 percent of the total population instead of over 20 percent, as our first simulation would have predicted.

The May 1986 change in behavior has, however, practically *no influence on the number of individuals falling ill in 1988*, nor could we expect it to have any significant impact given the long incubation period of AIDS. The number of people actually ill with AIDS on 31 December 1988 was 2,779. Our first simulation predicted 2,803 cases for that point in time if there was no change in behavior (fig. 24). With a change in behavior (the second simulation, described here), we predicted 2,709 cases.

The fact that a rather radical change in behavior in our model produced a barely perceptible change in the number of AIDS cases makes even more questionable the thesis that information and education have slowed the spread of the disease. We would hope that is true, and perhaps it is, but the drop in the number of cases to date offers no conclusive proof.

14. U. Reichert and D. Dörner, "Heurismen beim Umgang mit einem 'einfachen' dynamischen System," *Sprache und Kognition* 7 (1988): 12–24.

15. A. T. Bergerud, "Die Populationsdynamik von Räuber und Beute," *Spektrum der Wissenschaft*, part 2 (1984): 46–54.

16. W. Preussler, "Über die Bedingungen der Prognose eines bivariaten ökologischen Systems," University of Bamberg, 1985.

# 6. Planning

1. D. Dörner, *Die kognitive Organisation beim Problemlösen* (Bern: Huber, 1974), 137, 157.

2. K. Schütte, *Beweistheorie* (Berlin: Springer, 1960).

3. F. Klix, *Information und Verhalten* (Bern: Huber, 1971).

4. With his "morpological box" system, F. Zwicky (*Entdecken, Erfinden, Forschen im morphologischen Weltbild* [Munich: Droemer-Knaur, 1966]) created a formal structure for this method.

5. K. Duncker, *Zur Psychologie des produktiven Denkens* (Berlin: Springer, 1965).

6. H. Grote, *Bauen mit KOPF* (Berlin: Patzer, 1988), 65.

7. F. Malik, *Strategie des Managements komplexer Systeme* (St. Gallen: Institut für Betreibswirtschaft der Hochschule, 1985).

8. Schmerfeld, 54.

9. B. Brehmer and R. Allard, "Learning to Control a Dynamic System," *Learning and Instruction*, ed. E. de Corte et al. (Amsterdam: North-Holland, 1986).

10. Sunshine and Horowitz, 1968, cited in T. Roth, "Sprachstil und Problemlösekompetenz: Untersuchungen zum Formwortgebrauch im 'Lauten Denken' erfolgreicher und erfolgloser Bearbeiter 'komplexer' Probleme" (dissertation, University of Göttingen, 1986), 50.

11. Grote, 81–82.

12. Grote, 56.

13. T. Sarrazin, quoted in *Der Spiegel*, 28 March 1993.

14. J. D. Watson, *The Double Helix* (New York: Atheneum, 1968).

15. L. von Bertalanffy, *General System Theory* (New York: Braziller, 1968), 31.

16. A. Luchins and E. Luchins, "Mechanization in Problem-Solving: The Effect of *Einstellung*," *Psychological Monographs* 54, no. 6 (1942).

17. C. von Clausewitz, *On War*, ed. and trans. Michael Howard and Peter Paret (Princeton: Princeton University Press, 1984), 153, 154.

18. H. J. Kühle, "Zielangaben anstelle von Lösungen: Hintergründe für ein bei Politikern häufig zu beobachtendes Phänomen und dessen Konsequenzen," University of Bamberg, 1982; H. J. Kühle and P. Badke, "Die Entwicklung von Lösungsvorstellungen in komplexen Problemsituationen und die Gedächtnisstruktur," *Sprache und Kognition* 5, part 2 (1986): 95–105.

19. The correlation was 0.58.

20. The correlations were 0.64 and 0.67, respectively.

21. Kühle and Badke, "Die Entwicklung."

22. Roth, "Sprachstil und Problemelösekompetenz."

23. I. Kant, *Prolegomena to Any Future Metaphysics* (Indianapolis: Bobbs-Merrill, 1950), 10.

24. F. Reither, "Wertorientierung in komplexen Entscheidungssituationen," *Sprache und Kognition* 4, part 1 (1985), 21–27.

25. Reither measured variability as the relationship of the standard deviation of specific dosages to the average dosage.

26. The term is Stefan Strohschneider's.

## 7. So Now What Do We Do?

1. R. Kluwe, "Problemlösen, Entscheiden und Denkfehler," *Enzyklopädie der Psychologie: Ingenieurpsychologie*, ed. C. Hoyos and B. Zimolong (Göttingen: Hogrefe, 1988).

2. W. Putz-Osterloh, "Gibt es Experten für komplexe Probleme?" *Zeitschrift für Psychologie* 195 (1987): 63–84.

3. G. Vollmer, *Wissenschaft mit Steinzeitgehirnen?* Mannheimer Forum (Boehringer), 1986–87.

4. See P. Baltes et al., "One Facet of Successful Aging?" *Late Life Potential*, ed. M. Perlmutter (Washington, D.C.: Gerontological Society of America, 1988).

5. F. Reither, *Über die Selbstreflexion beim Problemlösen*, Term paper, Department of Psychology, University of Giessen, 1979.

6. C. Zangemeister, "Nutzwertanalyse von Projektalternativen," *Systemtheorie und Systemtechnik*, ed. F. Händle and F. Jensen (Munich: Nymphenburger, 1974).

7. D. E. Broadbent et al., "Implicit and Explicit Knowledge in the Control of Complex Systems," *British Journal of Psychology* 77 (1986): 33–50.

# Index

# About the Author

Dietrich Dörner is Professor of Psychology at the University of Bamberg. He is a winner of the Leibniz Prize, Germany's highest science award. An authority on cognitive behavior, his areas of specialty include logic and the theory of action. Dietrich Dörner is also director of the Cognitive Anthropology Project of the Max Planck Institute in Berlin.